AF682851

Matrix Structural Analysis

PUES-8

Pergamon Unified
Engineering Series

Pergamon Unified
Engineering Series

Matrix Structural Analysis

Jamal J. Azar
Dept. of Aerospace and Mechanical Engineering
University of Tulsa

PERGAMON PRESS INC.
NEW YORK · TORONTO · OXFORD
SYDNEY · BRAUNSCHWEIG

Pergamon Press Inc., Maxwell House, Fairview Park, Elmsford, New York 10523

Pergamon of Canada Ltd., 207 Queen's Quay West, Toronto 1

Pergamon Press Ltd., Headington Hill Hall, Oxford

Pergamon Press (Aust.) Pty. Ltd., 19a Boundary Street, Rushcutters Bay, N.S.W. 2011 Australia

Vieweg & Sohn GmbH, Burgplatz 1, Braunschweig

Copyright © 1972 Pergamon Press Inc.
Library of Congress Catalog Card No. 77-149553
First edition 1972

All rights reserved. No part of this publication may be reproduced, stored in a retrieval system, or transmitted in any form, or by any means, electronic, mechanical, photocopying, recording or otherwise, without the prior permission of Pergamon Press Inc.

Printed in the United States of America
08 016781 0 Hard cover

Pergamon Unified Engineering Series

GENERAL EDITORS

Thomas F. Irvine, Jr.
State University of New York at Stony Brook

James P. Hartnett
University of Illinois at Chicago Circle

EDITORS

William F. Hughes
Carnegie-Mellon University

Authur T. Murphy
PMC Colleges

William H. Davenport
Harvey Mudd College

Daniel Rosenthal
University of California, Los Angeles

SECTIONS

Continuous Media Section
Engineering Design Section
Engineering Systems Section
Humanities and Social Sciences Section
Information Dynamics Section
Materials Engineering Section
Engineering Laboratory Section

To Jean, Scott, Steven and my parents

Contents

Preface

With the advent of high-core-capacity, high-speed digital computers, matrix structural analysis has gained a great prominence throughout industry as a widely accepted method for the solution of highly complex structures, such as those found in space, sea and land-based vehicles.

Matrix structural analysis is predicated on the ability and skills of the analyst in obtaining a structural physical model whose behavior will predict closely the behavior of the actual system. Unless the real structure is conceived as having a finite number of degrees of freedom, the implementation of matrix techniques is not possible for practical solutions of structural problems. Thus, structural modeling or idealization is an invariable necessity in the field of matrix structural analysis.

In this book, the author attempts to present, in the simplest way, the theory and practical application of matrix structural analysis. The intent is to give the reader a basic understanding of modern sophisticated techniques in structural analysis, while keeping the mathematical treatment to a lesser degree of sophistication.

To the student, it offers a basic course in matrix structural analysis, at a senior or first year graduate level. To the practicing engineer who is familiar with the theory, it will serve as a quick reference where generalized stiffness matrices for various structural elements can be easily extracted and directly implemented into their own existing structural computer programs.

This book originated from the author's notes which have been used since 1966 in a formal three semester hour course taught at the University of Tulsa.

The method of presentation has proved to be readily comprehensible by the students concerned.

For those who are interested in graph theory and network technique in conjunction with matrix structural analysis, the author recommends the book by Dr. W. R. Spillers [24] of Columbia University.

The author wishes to express his appreciation to Dr. Heinz Bergmann, former professor at the University of Oklahoma, who first taught me the fundamentals of matrix structural analysis; to Mr. James A. Muckleroy, Senior Research Engineer, Pan American Petroleum, for his careful review of the manuscript and to my students at the University of Tulsa for their helpful comments. Appreciation is due also to Mrs. Maxine Lamb who typed the manuscript.

University of Tulsa
Oklahoma

J. J. AZAR

CHAPTER ONE

Matrix Algebra

1.1 INTRODUCTION

In this chapter the fundamental concepts of matrix algebra are introduced. Definitions and matrix manipulations are presented in a concise manner to provide adequate background for understanding the text.

1.2 DEFINITIONS

A *matrix* is an array of elements which can be written as follows:

$$[A] = \begin{bmatrix} A_{11} & A_{12} \cdots A_{1j} \\ A_{21} & A_{22} \cdots A_{2j} \\ \cdots & \cdots \quad \cdots \\ A_{i1} & A_{i2} \cdots A_{ij} \end{bmatrix} \qquad \begin{matrix} (i = 1, 2, 3, \ldots, m) \\ (j = 1, 2, 3, \ldots, n) \end{matrix} \tag{1.1}$$

where i and j indicate the number of rows and columns in the array respectively. Thus, the above matrix is referred to as a matrix of order $i \times j$. The A's are the elements which constitute the matrix. Their locations are determined by their subscripts. For instance, the element A_{i1} is in the ith row and column one; A_{1j} is in the first row and the jth column, and likewise, A_{ij} is in the ith row and jth column.

For $i \neq j$ as shown in Eq. (1.1), the matrix is referred to as a *rectangular matrix*. For $i = j$, the matrix is said to be a *square matrix*, i.e.:

$$[A] = \begin{bmatrix} A_{11} & A_{12} \cdots A_{1i} \\ A_{21} & A_{22} \cdots A_{2i} \\ \cdots & \cdots \quad \cdots \\ A_{i1} & A_{i2} \cdots A_{ii} \end{bmatrix} \tag{1.2}$$

The elements $A_{11}, A_{22}, \ldots, A_{ii}$ form the *main diagonal* of a square matrix.

For $i = 1$ and $j > 1$, the matrix is said to be a *row matrix*, i.e.,

$$\lfloor A \rfloor = [A_{11} \quad A_{12} \cdots A_{1j}] \qquad (j = 1, 2, \ldots, n) \tag{1.3}$$

For $i > 1$ and $j = 1$, the matrix is referred to as a *column matrix*, i.e.,

$$\{A\} = \begin{bmatrix} A_{11} \\ A_{21} \\ \cdots \\ \cdots \\ \cdots \\ A_{i1} \end{bmatrix} \qquad (i = 1, 2, \ldots, m) \tag{1.4}$$

Null, Diagonal and Unit Matrices

A *null matrix* is a matrix where elements are all zero, i.e.,

$$[A] \ \{A\} = \begin{bmatrix} 0 & 0 \cdots 0 \\ 0 & 0 \cdots 0 \\ \cdots & \cdots \ \cdots \\ 0 & 0 \cdots 0 \end{bmatrix} \tag{1.5}$$

The *diagonal matrix* is a square matrix where main diagonal elements, A_{ii}, are non-zero and its off-diagonal elements, $A_{ij} (i \neq j)$, are zero. Equation (1.6) illustrates such a matrix.

$$[A] = \begin{bmatrix} A_{11} & 0 & \cdots & 0 \\ 0 & A_{22} & \cdots & 0 \\ \cdots & \cdots & & \cdots \\ 0 & 0 & \cdots & A_{ii} \end{bmatrix} \tag{1.6}$$

A *unit matrix* or an *identity matrix* is a special case of a diagonal matrix where the main-diagonal elements, A_{ii}, are unity. The letter, I, symbolizes such a matrix.

$$[I] = \begin{bmatrix} 1 & 0 \cdots 0 \\ 0 & 1 \cdots 0 \\ \cdots & \cdots \ \cdots \\ 0 & 0 \cdots 1 \end{bmatrix} \tag{1.7}$$

Symmetric, Cross-Symmetric and Transposed Matrices

A *symmetric matrix* is a square matrix for which the following holds true;

$$A_{ij} = A_{ji}. \tag{1.8}$$

An illustrative example of a 4×4 square symmetric matrix is,

$$[A] = \begin{bmatrix} 3 & 4 & 2 & 1 \\ 4 & 5 & 6 & 8 \\ 2 & 6 & 7 & 9 \\ 1 & 8 & 9 & 6 \end{bmatrix} \tag{1.8a}$$

Notice in Eq. (1.8a) that the main diagonal along the dashed line is the line of symmetry.

The square matrix, $[A]$, is said to be cross-symmetric if the line of symmetry is the cross-diagonal, i.e.,

$$[A] = \begin{bmatrix} 1 & 2 & 3 & 4 \\ 7 & 8 & 5 & 3 \\ 0 & 6 & 8 & 2 \\ 9 & 0 & 7 & 1 \end{bmatrix} \tag{1.8b}$$

The *transpose*, $[A]^T$, of a matrix $[A]$ is found by writing the A_{ij} elements of $[A]$ as the A_{ji} elements of $[A]^T$. The example below gives a numerical illustration of matrix transposing procedure.

If

$$[A] = \begin{bmatrix} 1 & 2 & 4 & 6 \\ 5 & 8 & 9 & 3 \\ 4 & 7 & 2 & 4 \end{bmatrix} \tag{1.8c}$$

then

$$[A]^T = \begin{bmatrix} 1 & 5 & 4 \\ 2 & 8 & 7 \\ 4 & 9 & 2 \\ 6 & 3 & 4 \end{bmatrix} \tag{1.8d}$$

The transpose of a square symmetric matrix, $[A]$, is the matrix itself, i.e.,

$$[A]^T = [A] \tag{1.9}$$

Determinants are defined only for square matrices. Thus, for a given square matrix $[A]$, the determinant is defined as the number which results upon performing the following arithmetic operation on $[A]$,

$$|A| = \sum_{i=1}^{n} A_{ij} C_{ij} \qquad (j = 1, 2, \ldots, n) \tag{1.10}$$

or

$$|A| = \sum_{j=1}^{n} A_{ij} C_{ij} \qquad (i = 1, 2, \ldots, n) \tag{1.11}$$

The elements, A_{ij} in Eqs. (1.10) and (1.11) are the jth column and ith row elements of the matrix, $[A]$, respectively. Likewise, C_{ij} are the cofactors corresponding to the jth column (Eq. 1.10) and ith row (Eq. 1.11) of matrix $[A]$.

Mathematically, the co-factors of a matrix, $[A]$, are defined as follows:

$$C_{ij} = (-1)^{i+j} M_{ij}. \tag{1.12}$$

Where M_{ij} are the minors of matrix $[A]$, and are defined as the determinant of the matrix which results after the ith row and the jth column are deleted from matrix, $[A]$.

Equations (1.10), (1.11), and (1.12) can be best understood by considering the following numerical example.

Given the following 3×3 square matrix, it is required to find its determinant.

$$[A] = \begin{bmatrix} 5 & 6 & 2 \\ 3 & 4 & 2 \\ 4 & 0 & 3 \end{bmatrix} \tag{1.12a}$$

Let us first apply Eq. (1.10) and arbitrarily pick $j = 2$. Hence Eq. (1.10) can be written as,

$$|A| = \sum_{i=1}^{3} A_{i2} C_{i2} = A_{12} C_{12} + A_{22} C_{22} + A_{32} C_{32}.$$

From Eq. (1.12a) column 2 of matrix, $[A]$,

$$A_{12} = 6, \qquad A_{22} = 4, \qquad A_{32} = 0.$$

For $j = 2$, Eq. (1.12) becomes

$$C_{i2} = (-1)^{i+2} M_{i2} \qquad (i = 1, 2, 3)$$

therefore

$$C_{12} = -M_{12}, \qquad C_{22} = M_{22}, \qquad C_{32} = -M_{32}.$$

By definition, the minor, M_{12}, of matrix $[A]$, is the determinant of that matrix after deleting the first row and the second column. Or

$$M_{12} = \begin{vmatrix} 3 & 2 \\ 4 & 3 \end{vmatrix} = (3 \times 3) - (4 \times 2) = 9 - 8 = 1,$$

likewise

$$M_{22} = \begin{vmatrix} 5 & 2 \\ 4 & 3 \end{vmatrix} = (5 \times 3) - (4 \times 2) = 7,$$

and

$$M_{32} = \begin{vmatrix} 5 & 2 \\ 3 & 2 \end{vmatrix} = (5 \times 2) - (3 \times 2) = 4,$$

hence

$$C_{12} = -1, \qquad C_{22} = 7, \qquad C_{32} = -4,$$

therefore the determinant is

$$|A| = (6)(-1) + (5)(7) + (0)(-4) = 29.$$

Now applying Eq. (1.11) instead and arbitrarily choosing $i = 3$, yields

$$|A| = \sum_{j=1}^{3} A_{3j}C_{3j} = A_{31}C_{31} + A_{32}C_{32} + A_{33}C_{33}.$$

From Eq. (1.12a) the third row of matrix $[A]$ is

$$A_{31} = 4, \qquad A_{32} = 0, \qquad A_{33} = 3.$$

From Eq. (1.12) for $i = 3$ we have,

$$C_{3j} = (-1)^{3+j}M_{3j} \qquad (j = 1, 2, 3)$$

or

$$C_{31} = M_{31}, \qquad C_{32} = -M_{32}, \qquad C_{33} = M_{33},$$

where again,

$$M_{31} = \begin{vmatrix} 6 & 2 \\ 5 & 2 \end{vmatrix} = 2, \qquad M_{32} = \begin{vmatrix} 5 & 2 \\ 3 & 2 \end{vmatrix} = 4,$$

and

$$M_{33} = \begin{vmatrix} 5 & 6 \\ 3 & 5 \end{vmatrix} = 7,$$

hence

$$|A| = (4)(2) + (0)(4) + (3)(7) = 29,$$

which is the same result obtained previously.

Rules

a. If the elements of any row or column of $|A|$ has a common factor, b, then the following holds true

$$\begin{vmatrix} A_{11}b & A_{12}b \\ A_{21} & A_{22} \end{vmatrix} = b\begin{vmatrix} A_{11} & A_{12} \\ A_{21} & A_{22} \end{vmatrix},$$

or

$$\begin{vmatrix} A_{11}b & A_{12} \\ A_{21}b & A_{22} \end{vmatrix} = b\begin{vmatrix} A_{11} & A_{12} \\ A_{21} & A_{22} \end{vmatrix},$$

and

$$\begin{vmatrix} A_{11}b & A_{12}b \\ A_{21}b & A_{22}b \end{vmatrix} = b^2\begin{vmatrix} A_{11} & A_{12} \\ A_{21} & A_{22} \end{vmatrix}.$$

The converse of this argument is that having $|A|$ multiplied by a factor is equivalent to multiplying any row or column of $|A|$ by the same factor.

b. When rows and corresponding columns in $|A|$ are interchanged the value of the determinant of $|A|$ is not altered, i.e.,

$$\begin{vmatrix} 2 & 4 \\ 5 & 6 \end{vmatrix} = \begin{vmatrix} 2 & 5 \\ 4 & 6 \end{vmatrix} = -8.$$

c. When two rows or two columns of $|A|$ are interchanged the sign of the determinant is changed, i.e.,

$$\begin{vmatrix} 2 & 4 \\ 5 & 6 \end{vmatrix} = -\underset{\substack{\text{Column}\\\text{Interchange}}}{\begin{vmatrix} 4 & 2 \\ 6 & 5 \end{vmatrix}} = -\underset{\substack{\text{Row}\\\text{Interchange}}}{\begin{vmatrix} 5 & 6 \\ 2 & 4 \end{vmatrix}} = -8.$$

d. If two rows or two columns of $|A|$ are identical then the value of $|A|$ is zero.

e. If a row or a column is changed by adding to or subtracting from its elements the corresponding elements (or the corresponding elements multiplied by a common factor) of any other row or column, the value of $|A|$ is not altered.

Examples

$$\begin{vmatrix} 2 & 4 \\ 5 & 6 \end{vmatrix} = \overset{\substack{\text{Row 2 added}\\\text{to row 1}}}{\begin{vmatrix} 7 & 10 \\ 5 & 6 \end{vmatrix}} = \overset{\substack{\text{Column 1 added}\\\text{to column 2}}}{\begin{vmatrix} 2 & 6 \\ 5 & 11 \end{vmatrix}} = \underset{\substack{\text{Row 1 mult.}\\\text{by } \frac{1}{2} \text{ and}\\\text{result added}\\\text{to row 2.}}}{\begin{vmatrix} 2 & 4 \\ 6 & 8 \end{vmatrix}} = \underset{\substack{\text{Column 2 mult.}\\\text{by } \frac{1}{2} \text{ and}\\\text{result added}\\\text{to column 1.}}}{\begin{vmatrix} 4 & 4 \\ 8 & 6 \end{vmatrix}} = -8$$

Performing the same steps as above but subtracting instead of adding yields

$$\begin{vmatrix} 2 & 4 \\ 5 & 6 \end{vmatrix} = \begin{vmatrix} -3 & 2 \\ 5 & 6 \end{vmatrix} = \begin{vmatrix} 2 & 2 \\ 5 & 1 \end{vmatrix} = \begin{vmatrix} 2 & 4 \\ 4 & 4 \end{vmatrix} = \begin{vmatrix} 2 & 4 \\ 5 & 6 \end{vmatrix} = -8$$

f. If the determinant, $|A|$, has any dependent rows or columns its value is zero.

Example

$$|A| = \begin{vmatrix} 1 & 2 & 3 \\ 4 & 6 & 11 \\ 2 & 8 & 8 \end{vmatrix} = 0$$

Column 3 is dependent on columns 1 and 2, i.e.,
Column 3 = 2 (column 1) + $\frac{1}{2}$ (column 2).

Singular Matrices

The matrix, $[A]$, is said to be singular if the value of its determinant is zero.

Example

If

$$[A] = \begin{bmatrix} 1 & 2 \\ 2 & 4 \end{bmatrix},$$

then the determinant of $[A]$ is,

$$|A| = \begin{vmatrix} 1 & 2 \\ 2 & 4 \end{vmatrix} = 0,$$

hence $[A]$ is singular.

1.3 MATRIX ALGEBRA MANIPULATION

Matrix Addition

Two matrices $[A]^{i\times j}$ and $[B]^{i\times j}$, can be added by adding each element A_{ij} of Matrix $[A]$, to the corresponding element, B_{ij}, of Matrix $[B]$.

$$\begin{bmatrix} A_{11} & A_{12} \\ A_{21} & A_{22} \end{bmatrix} + \begin{bmatrix} B_{11} & B_{12} \\ B_{21} & B_{22} \end{bmatrix} = \begin{bmatrix} (A_{11}+B_{11}) & (A_{12}+B_{12}) \\ (A_{21}+B_{21}) & (A_{22}+B_{22}) \end{bmatrix}, \qquad (1.13)$$

or

$$[A]^{i\times j} + [B]^{i\times j} = [C]^{i\times j}, \qquad (1.14)$$

where C_{ij} in general is given by,

$$C_{ij} = A_{ij} + B_{ij}. \qquad (1.15)$$

Matrix Subtraction

Matrix subtraction is analogous to matrix addition where the elements A_{ij} and B_{ij} are subtracted instead of being added.

$$[A]^{i\times j} - [B]^{i\times j} = [D]^{i\times j}, \qquad (1.16)$$

where

$$D_{ij} = A_{ij} - B_{ij}. \qquad (1.17)$$

Matrix Multiplication

Multiplication of two matrices, $[A]^{i\times j}$ and $[B]^{j\times s}$ is performed by multiplying each element of row m of $[A]^{i\times j}$, by its corresponding element in column n of $[B]^{j\times s}$ and the products summed.

Mathematically this can be expressed as follows,

$$[A]^{i\times j}[B]^{j\times s}=[C]^{i\times s}. \tag{1.18}$$

where an element C_{mn} of $[C]^{i\times s}$ is given by,

$$C_{mn}=\sum_{k=1}^{j} A_{mk}B_{kn} \qquad (j=\text{number of columns in } A \text{ or rows in } B.) \tag{1.19}$$

schematically this is shown as follows,

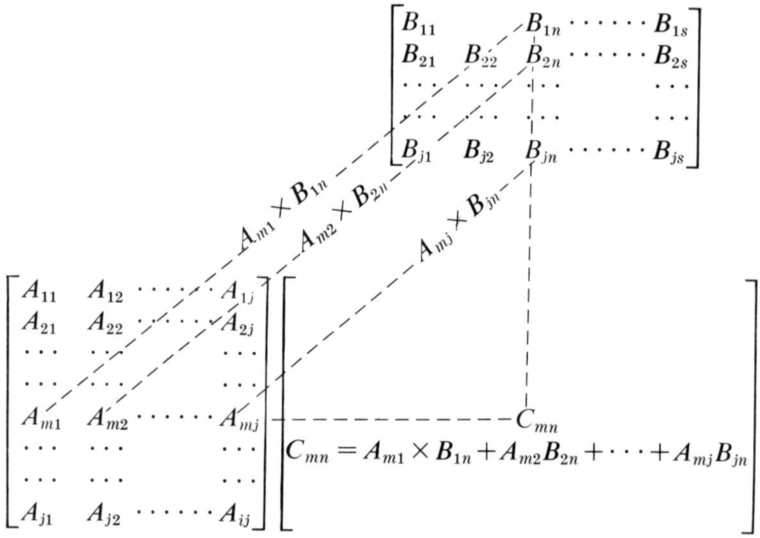

(1.20)

Numerical example

Given

$$[A]=\begin{bmatrix}1 & 3\\5 & 4\end{bmatrix}$$

and

$$[B]=\begin{bmatrix}3 & 5 & 6\\1 & 2 & 3\end{bmatrix}$$

Required

$$[A][B]=[C]$$

Solution

$$\begin{bmatrix}1 & 3\\5 & 4\end{bmatrix}\begin{bmatrix}3 & 5 & 6\\1 & 2 & 3\end{bmatrix}=\begin{bmatrix}C_{11} & C_{12} & C_{13}\\C_{21} & C_{22} & C_{23}\end{bmatrix}$$

from Eq. (1.19) we get

$$C_{mn} = \sum_{k=1}^{2} A_{mk}B_{kn} = A_{m1}B_{1n} + A_{m2}B_{2n}$$

$$C_{11} = A_{11}B_{11} + A_{12}B_{21} = 1\times 3 + 3\times 1 = 6$$

$$C_{12} = A_{11}B_{12} + A_{12}B_{22} = 1\times 5 + 3\times 2 = 11$$

$$C_{13} = A_{11}B_{13} + A_{12}B_{23} = 1\times 6 + 3\times 3 = 15$$

$$C_{21} = A_{21}B_{11} + A_{22}B_{21} = 5\times 3 + 4\times 1 = 19$$

$$C_{22} = A_{21}B_{12} + A_{22}B_{22} = 5\times 5 + 4\times 2 = 33$$

$$C_{23} = A_{21}B_{13} + A_{22}B_{23} = 5\times 6 + 4\times 3 = 42$$

Note:

Two matrices cannot be multiplied unless the number of rows of one matrix is the same as the number of columns in the second, i.e.,

$$[A]^{i\times j}[B]^{r\times s} = [C]^{i\times s},$$

only if $j = r$, otherwise the multiplication cannot be performed.

Matrix Multiplication Properties

1. Associative Law

$$[D] = ([A][B])[C] = [A]([B][C])$$

and in general any element of $[D]$ is given by,

$$D_{ij} = \sum_r \sum_s A_{ir}B_{rs}C_{sj}. \tag{1.21}$$

2. Distributive Law

$$[D] = [A]([B]+[C]) = [A][B]+[A][C].$$

3. Commutative Law

$$[A][B] \neq [B][A].$$

4. If $[A]$ and $[B]$ are both symmetric matrices their product, generally, does not yield a symmetric matrix.

5. The matrix, $[A]$, if multiplied or divided by a scalar quantity, b, is equivalent to multiplying or dividing each element of $[A]$ by b.

Example

$$2\times\begin{bmatrix}2 & 4\\ 5 & 10\end{bmatrix} = \begin{bmatrix}2\times 2 & 2\times 4\\ 2\times 5 & 2\times 10\end{bmatrix}$$

or

$$\tfrac{1}{2} \times \begin{bmatrix} 2 & 4 \\ 4 & 10 \end{bmatrix} = \begin{bmatrix} \frac{2}{2} & \frac{4}{2} \\ \frac{5}{2} & \frac{10}{2} \end{bmatrix}$$

Matrix Inversion

The inverse of a matrix, $[A]$, is denoted by $[A]^{-1}$ and defined as,

$$[A]^{-1}[A] = [A][A]^{-1} = [I]. \qquad (1.22)$$

Only square nonsingular matrices have an inverse.

The inversion of a square nonsingular matrix can be obtained by using Eqs. (1.10) or (1.11) and (1.12).

The inversion of matrix $[A]$, can be accomplished as follows:

Step 1. Form a new matrix of co-factors C_{ij} by using Eq. (1.12). This new matrix of co-factors will be referred to as the adj $[A]$ of $[A]$.

$$C_{ij} = (-1)^{i+j} M_{ij} \qquad (1.12)$$

Step 2. Transpose adj $[A]$ and call the result adj $[A]^T$. Reference to Eq. (1.9) might be helpful.

Step 3. Evaluate the determinant of $[A]$. This can be accomplished by using Eqs. (1.10) or (1.11).

$$|A| = \sum_{i=1}^{n} A_{ij} C_{ij} \qquad (j = 1, 2, \ldots, n) \qquad (1.10)$$

or

$$|A| = \sum_{j=1}^{n} A_{ij} C_{ij} \qquad (i = 1, 2, \ldots, n) \qquad (1.11)$$

Step 4. Divide adj $[A]^T$ by $|A|$ to obtain the inverse, $[A]^{-1}$, or

$$[A]^{-1} = \frac{\text{adj } [A]^T}{|A|} \qquad (1.23)$$

Example

Given a 3×3 matrix as shown below, find its inverse.

$$[A] = \begin{bmatrix} 5 & 6 & 2 \\ 3 & 5 & 2 \\ 4 & 0 & 3 \end{bmatrix}$$

Step 1. Find the adj $[A]$. This is done by constructing a new matrix of co-factors of $[A]$.

From Eq. (1.12)

$$C_{ij} = (-1)^{i+j} M_{ij}$$

or

$$C_{11} = M_{11}, \quad C_{12} = -M_{12}, \quad C_{13} = M_{13},$$
$$C_{21} = -M_{21}, \quad C_{22} = M_{22}, \quad C_{23} = -M_{23},$$
$$C_{31} = M_{31}, \quad C_{32} = -M_{32}, \quad C_{33} = M_{33},$$

where

$$M_{11} = \begin{vmatrix} 5 & 2 \\ 0 & 3 \end{vmatrix} = 15, \qquad M_{12} = \begin{vmatrix} 3 & 2 \\ 4 & 3 \end{vmatrix} = 1$$

$$M_{13} = \begin{vmatrix} 3 & 5 \\ 4 & 0 \end{vmatrix} = -20, \qquad M_{21} = \begin{vmatrix} 6 & 2 \\ 0 & 3 \end{vmatrix} = 18$$

$$M_{22} = \begin{vmatrix} 5 & 2 \\ 4 & 3 \end{vmatrix} = 7, \qquad M_{23} = \begin{vmatrix} 5 & 6 \\ 4 & 0 \end{vmatrix} = -24$$

$$M_{33} = \begin{vmatrix} 5 & 6 \\ 3 & 5 \end{vmatrix} = 7, \qquad M_{31} = \begin{vmatrix} 6 & 2 \\ 5 & 2 \end{vmatrix} = 2$$

$$M_{32} = \begin{vmatrix} 5 & 2 \\ 3 & 2 \end{vmatrix} = 4$$

therefore

$$\text{adj } [A] = \begin{bmatrix} 15 & -1 & -20 \\ -18 & 7 & 24 \\ 2 & -4 & 7 \end{bmatrix}$$

Step 2. Find the transpose of adj $[A]$.

$$\text{adj } [A]^T = \begin{bmatrix} 15 & -18 & 2 \\ -1 & 7 & -4 \\ -20 & 24 & 7 \end{bmatrix}$$

Step 3. Find the determinant of $[A]$.
From Eq. (1.11) for $j = 2$,

$$\begin{aligned} |A| &= A_{12}C_{12} + A_{22}C_{22} + A_{32}C_{32} \\ &= 6(-1) + 5(7) + 0(-4) = 29 \end{aligned}$$

Step 4. Divide adj $[A]^T$ by $|A|$, (Eq. 1.23),

$$[A]^{-1} = \frac{\text{adj } [A]^T}{|A|} = \frac{1}{29}\begin{bmatrix} 15 & -18 & 2 \\ -1 & 7 & -4 \\ -20 & 24 & 7 \end{bmatrix}$$

Check

$$[A][A]^{-1} = [I]$$

$$\begin{bmatrix} 5 & 6 & 2 \\ 3 & 5 & 2 \\ 4 & 0 & 3 \end{bmatrix} \begin{bmatrix} \frac{15}{29} & -\frac{18}{29} & \frac{2}{29} \\ -\frac{1}{29} & \frac{7}{29} & -\frac{4}{29} \\ -\frac{20}{29} & \frac{24}{29} & \frac{7}{29} \end{bmatrix} = \begin{bmatrix} 1 & 0 & 0 \\ 0 & 1 & 0 \\ 0 & 0 & 1 \end{bmatrix}$$

Since an identity matrix is obtained the inversion is correct.

The *LU* Method for Matrix Inversion

The *LU* method of matrix inversion (Cholesky method) is based on the fact that any given nonsingular square matrix can be decomposed into a lower triangular matrix $[L]$ and an upper triangular matrix $[U]$. Thus, if $[S]$ is a nonsingular matrix, it can be decomposed as follows,

$$[S] = [L][U]. \tag{1.24}$$

Restricting ourselves to symmetrical matrices only, in which case $[U]$ is the transpose of $[L]$, Eq. (1.24) then becomes

$$[S] = [L][L]^T, \tag{1.24a}$$

or, in an expanded form

$$\begin{bmatrix} S_{11} & S_{12} & S_{13} \cdots S_{1n} \\ S_{21} & S_{22} & S_{23} \cdots S_{2n} \\ S_{31} & S_{32} & S_{33} \cdots S_{3n} \\ \cdots & \cdots & \cdots \quad \cdots \\ S_{n1} & S_{n2} & S_{n3} \cdots S_{nn} \end{bmatrix} = \begin{bmatrix} L_{11} & 0 & 0 \cdots 0 \\ L_{21} & L_{22} & 0 \cdots 0 \\ L_{31} & L_{32} & L_{33} \cdots 0 \\ \cdots & \cdots & \cdots \quad \cdots \\ L_{n1} & L_{n2} & L_{n3} \cdots L_{nn} \end{bmatrix} \begin{bmatrix} L_{11} & L_{21} & L_{31} \cdots L_{n1} \\ 0 & L_{22} & L_{32} \cdots L_{n2} \\ 0 & 0 & L_{33} \cdots L_{n3} \\ \cdots & \cdots & \cdots \quad \cdots \\ 0 & 0 & 0 \cdots L_{nn} \end{bmatrix} \tag{1.25}$$

If matrix multiplication is carried out on the righthand side of Eq. (1.25), then equating the coefficients of the matrices on both sides of the equation yields,

$$\begin{aligned} S_{11} &= L_{11}^2 \\ S_{22} &= L_{21}^2 + L_{22}^2 \\ S_{33} &= L_{31}^2 + L_{32}^2 + L_{33}^2 \\ \cdots \\ S_{nn} &= L_{n1}^2 + L_{n2}^2 + L_{n3}^2 + \cdots + L_{nn}^2. \end{aligned}$$

Or in general for $(i = j)$,

$$S_{ii} = \sum_{m=1}^{i} L_{im}^2 \qquad (i = j) \tag{1.26}$$

also,

$$S_{21} = L_{11}L_{21}$$
$$S_{31} = L_{11}L_{31}$$
$$\cdots$$
$$S_{n1} = L_{11}L_{n1}$$
$$S_{32} = L_{21}L_{31} + L_{22}L_{32}$$
$$\cdots$$
$$S_{n3} = L_{n1}L_{31} + L_{n2}L_{32} + L_{n3}L_{33}.$$

Or in general for $(i > j)$,

$$S_{ij} = \sum_{m=1}^{j} L_{im}L_{jm} \qquad (i > j). \tag{1.27}$$

Equations (1.26) and (1.27) may be rewritten as follows,

$$S_{ii} = L_{ii}^2 + \sum_{m=1}^{i-1} L_{im}^2 \qquad (j = i > 1) \tag{1.26a}$$

and

$$S_{ij} = L_{jj}L_{ij} + \sum_{m=1}^{j-1} L_{im}L_{jm} \qquad (i > 1, j) \tag{1.27a}$$

Hence the coefficients of the upper triangular matrix $[L]$ can be expressed as,

$$L_{ii} = \left(S_{ii} - \sum_{m=1}^{i-1} L_{im}^2\right)^{1/2} \qquad (i = j) \tag{1.28}$$

$$L_{ij} = \frac{S_{ij} - \sum_{m=1}^{j-1} L_{im}L_{jm}}{L_{jj}} \qquad (i > j) \tag{1.29}$$

$$L_{ij} = 0 \qquad (i < j) \tag{1.30}$$

Having formed the lower triangular matrix $[L]$, we shall now show how the inverse of $[A]$ can be obtained working on $[L]$ only. From Eq. (1.24a),

$$[S]^{-1} = [[L][L^T]]^{-1} = [L^T]^{-1}[L]^{-1} = [L^{-1}]^T[L]^{-1} \tag{1.31}$$

If we let

$$[R] = [L]^{-1},$$

then Eq. (1.31) becomes

$$[S]^{-1} = [R]^T[R] \tag{1.32}$$

To find the coefficients of $[R]$ we shall utilize the fact that the inverse of a matrix if premultiplied by the matrix itself yields an identity matrix $[I]$, i.e.,

$$[R][L] = [I], \tag{1.33}$$

or in an expanded form, Eq. (1.33) is,

$$\begin{bmatrix} R_{11} & 0 & 0 & \cdots & 0 \\ R_{21} & R_{22} & 0 & \cdots & 0 \\ R_{31} & R_{32} & R_{33} & \cdots & 0 \\ \cdots & \cdots & \cdots & & \cdots \\ R_{n1} & R_{n2} & R_{n3} & \cdots & R_{nn} \end{bmatrix} \begin{bmatrix} L_{11} & 0 & 0 & \cdots & 0 \\ L_{21} & L_{22} & 0 & \cdots & 0 \\ L_{31} & L_{32} & L_{33} & \cdots & 0 \\ \cdots & \cdots & \cdots & & \cdots \\ L_{n1} & L_{n2} & L_{n3} & \cdots & L_{nn} \end{bmatrix} = \begin{bmatrix} 1 & 0 & 0 \cdots 0 \\ 0 & 1 & 0 \cdots 0 \\ 0 & 0 & 1 \cdots 0 \\ \cdots & \cdots & \cdots \\ 0 & 0 & 0 \cdots 1 \end{bmatrix} \tag{1.33a}$$

Performing the matrix multiplication on the left hand side of Eq. (1.33a), equating the coefficients to those on the right hand side and solving for R_{ij} yields,

$$R_{ii} = 1/L_{ii} \qquad (i = j) \tag{1.34}$$

$$R_{ij} = -\frac{\sum_{m=j+1}^{i} R_{im}L_{mj}}{L_{jj}} \qquad (i > j) \tag{1.35}$$

$$R_{ij} = 0 \qquad (i < j) \tag{1.36}$$

From Eq. (1.32) the inverse of the matrix $[S]$ is,

$$[S]^{-1} = [H] = [R]^T[R], \tag{1.37}$$

in which the coefficients of $[H]$ are expressed by,

$$H_{ij} = \sum_{m=j}^{n} R_{im}R_{mi} \qquad (i \geqslant j) \tag{1.38}$$

$$H_{ij} = H_{ji}. \tag{1.39}$$

Example

Use the LU method to find the inverse of the following matrix

$$[S] = \begin{bmatrix} 1 & 2 & 1 \\ 2 & 3 & 2 \\ 1 & 2 & 2 \end{bmatrix}.$$

Forming the $[L]$ matrix from Eqs. (1.28) and (1.29) yields,

$$L_{11} = \sqrt{S_{11}} = 1$$

$$L_{21} = S_{21}/L_{11} = 2$$

$$L_{31} = S_{31}/L_{11} = 1$$

$$L_{22} = \left(S_{22} - \sum_{m=1}^{1} L_{2m}\right)^{1/2} = (3-4)^{1/2} = \sqrt{-1}$$

$$L_{32} = \left(S_{22} - \sum_{m=1}^{1} L_{3m}L_{2m}\right)/L_{22} = 0$$

$$L_{33} = \left(S_{33} - \sum_{m=1}^{2} L_{3m}\right) = 1$$

$$L_{ij} = 0 \qquad (i < j)$$

The elements of the $[R]$ matrix from Eqs. (1.34) and (1.35) are

$$R_{33} = \frac{1}{L_{33}} = 1, \qquad R_{22} = \frac{1}{L_{22}} = 1/\sqrt{-1}, \qquad R_{11} = \frac{1}{L_{11}} = 1$$

$$R_{32} = -\left(\sum_{m=3}^{3} R_{3m}L_{m2}\right)/L_{22} = 0$$

$$R_{31} = -\left(\sum_{m=2}^{3} R_{3m}L_{m1}\right)/L_{11} = -1$$

$$R_{21} = -\left(\sum_{2}^{2} R_{2m}L_{m1}\right)/L_{11} = -2/\sqrt{-1}$$

or

$$[R] = \begin{bmatrix} 1 & 0 & 0 \\ 2\sqrt{-1} & -\sqrt{-1} & 0 \\ -1 & 0 & 1 \end{bmatrix}$$

$$[R]^T = \begin{bmatrix} 1 & 2\sqrt{-1} & -1 \\ 0 & -\sqrt{-1} & 0 \\ 0 & 0 & 1 \end{bmatrix}$$

therefore, the inverse of the matrix $[S]$, from Eq. (1.32) is

$$[S]^{-1} = [R]^T[R] = \begin{bmatrix} 1 & 2\sqrt{-1} & -1 \\ 0 & -\sqrt{-1} & 0 \\ 0 & 0 & 1 \end{bmatrix} \begin{bmatrix} 1 & 0 & 0 \\ 2\sqrt{-1} & -\sqrt{-1} & 0 \\ -1 & 0 & 1 \end{bmatrix}$$

$$= \begin{bmatrix} -2 & 2 & -1 \\ 2 & -1 & 0 \\ -1 & 0 & 1 \end{bmatrix}$$

Properties of the Inverse Matrix

1. The inverse of a symmetric matrix is also symmetric.
2. $[AB]^{-1} = [B]^{-1}[A]^{-1}$. (1.40)

Matrix Partitioning

It is often convenient to partition a matrix into smaller submatrices. This can be accomplished by drawing vertical and horizontal dotted lines as shown below

$$[A] = \left[\begin{array}{cc|cc} A_{11} & A_{12} & A_{13} & A_{14} \\ A_{21} & A_{22} & A_{23} & A_{24} \\ \hline A_{31} & A_{32} & A_{33} & A_{34} \\ A_{41} & A_{42} & A_{43} & A_{44} \end{array}\right]$$

or in shorthand matrix notation

$$[A] = \left[\begin{array}{c|c} A_{\alpha\alpha} & A_{\alpha\beta} \\ \hline A_{\beta\alpha} & A_{\beta\beta} \end{array}\right]$$

$$[B] = \left[\begin{array}{c} B_{11} \\ B_{21} \\ \hline B_{31} \\ B_{41} \\ B_{51} \end{array}\right]$$

$$[C] = \left[\begin{array}{c|cc} C_{11} & C_{12} & C_{13} \\ \hline C_{21} & C_{22} & C_{23} \\ C_{31} & C_{32} & C_{33} \end{array}\right]$$

Properties of Partitioned Matrices

1. Two partitioned matrices, $[A]^{i\times j}$ and $[B]^{i\times j}$ can be added or subtracted in terms of their submatrices if and only if $[A]$ and $[B]$ are of the same order and partitioned in the same manner.

Example

$$[A] = \left[\begin{array}{cc|c} 1 & 2 & 3 \\ \hline 4 & 5 & 6 \\ 7 & 8 & 9 \end{array}\right] \qquad [B] = \left[\begin{array}{cc|c} 3 & 4 & 5 \\ \hline 2 & 1 & 8 \\ 6 & 7 & 9 \end{array}\right]$$

$$[A] + [B] = [C]$$

Before partitioning this can be done by adding each element A_{ij} to the corresponding element B_{ij}, or

$$C = \begin{bmatrix} 4 & 6 & 8 \\ 6 & 6 & 14 \\ 13 & 15 & 18 \end{bmatrix}$$

After partitioning the addition of the two matrices in terms of the submatrices can be carried as follows

$$[C] = \left[\begin{array}{cc|c} 1 & 2 & 3 \\ \hline 4 & 5 & 6 \\ 7 & 8 & 9 \end{array}\right] + \left[\begin{array}{cc|c} 3 & 4 & 5 \\ \hline 2 & 1 & 8 \\ 6 & 7 & 9 \end{array}\right] = \begin{bmatrix} [A_{11}] + [B_{11}] & [A_{12}] + [B_{12}] \\ [A_{21}] + [B_{21}] & [A_{22}] + [B_{22}] \end{bmatrix}$$

where

$$[A_{11}] = [1 \quad 2]$$
$$[B_{11}] = [3 \quad 4]$$
$$A_{12} = 3$$
$$B_{12} = 5$$
$$[A_{21}] = \begin{bmatrix} 4 & 5 \\ 7 & 8 \end{bmatrix}, \qquad B_{21} = \begin{bmatrix} 2 & 1 \\ 6 & 7 \end{bmatrix}$$
$$[A_{22}] = \begin{bmatrix} 6 \\ 9 \end{bmatrix}, \qquad B_{22} = \begin{bmatrix} 8 \\ 9 \end{bmatrix}$$

therefore

$$[C] = \begin{bmatrix} 4 & 6 & 8 \\ 6 & 6 & 14 \\ 13 & 15 & 18 \end{bmatrix}$$

In general, two matrices, $[A]$ and $[B]$ can be added in terms of their partitioned matrices as

$$\left[\begin{array}{c|c} A_{\alpha\alpha} & A_{\alpha\beta} \\ \hline A_{\beta\alpha} & A_{\beta\beta} \end{array}\right] + \left[\begin{array}{c|c} B_{\alpha\alpha} & B_{\alpha\beta} \\ \hline B_{\beta\alpha} & B_{\beta\beta} \end{array}\right] = \left[\begin{array}{c|c} A_{\alpha\alpha} + B_{\alpha\alpha} & A_{\alpha\beta} + B_{\alpha\beta} \\ \hline A_{\beta\alpha} \quad B_{\beta\alpha} & A_{\beta\beta} + B_{\beta\beta} \end{array}\right] \tag{1.41}$$

Subtraction can be carried out in the same manner as for the addition.

2. Two partitioned matrices, $[A]^{m\times n}$ and $[B]^{n\times q}$ can be multiplied in terms of their submatrices if and only if the submatrices of $[A]$ and the corresponding submatrices of $[B]$ obey the law of matrix multiplication, i.e., if $[A_a]^{r\times s}$ is a submatrix of $[A]$ and $[B_b]^{p\times q}$ is a submatrix of $[B]$ then multiplication is only possible if $p = s$.

Example

$$[A]^{3\times 3} = \begin{array}{c} \quad 2\times 2 \qquad 2\times 1 \\ \left[\begin{array}{cc|c} 1 & 2 & 3 \\ 4 & 5 & 6 \\ \hline 7 & 8 & 9 \end{array}\right] \\ \quad 1\times 2 \qquad 1\times 1 \end{array}$$

$$[B]^{3\times 2} = \begin{array}{c} 2\times 1 \quad 2\times 1 \\ \left[\begin{array}{c|c} 3 & 5 \\ 1 & 3 \\ \hline 4 & 2 \end{array}\right] \\ 1\times 1 \quad 1\times 1 \end{array}$$

$$[A][B]=\left[\begin{array}{cc|c}1&2&3\\4&5&6\\\hline 7&8&9\end{array}\right]\left[\begin{array}{c|c}3&5\\1&3\\\hline 4&2\end{array}\right]$$

$$=\left[\begin{array}{c|c}\begin{bmatrix}1&2\\4&5\end{bmatrix}\times\begin{bmatrix}3\\1\end{bmatrix}+\begin{bmatrix}3\\6\end{bmatrix}\times 4 & \begin{bmatrix}1&2\\4&5\end{bmatrix}\times\begin{bmatrix}5\\3\end{bmatrix}+\begin{bmatrix}3\\6\end{bmatrix}\times 2\\\hline [7\quad 8]\times\begin{bmatrix}3\\1\end{bmatrix}+9\times 4 & [7\quad 8]\times\begin{bmatrix}5\\3\end{bmatrix}+9\times 2\end{array}\right]$$

Performing the multiplication of the inner submatrices yields

$$[A][B]=\begin{bmatrix}17&17\\41&47\\65&77\end{bmatrix}$$

In general

$$\left[\begin{array}{c|c}[A_{\alpha\alpha}]&[A_{\alpha\beta}]\\\hline [A_{\beta\alpha}]&[A_{\beta\beta}]\end{array}\right]\left[\begin{array}{c|c}[B_{\alpha\alpha}]&[B_{\alpha\beta}]\\\hline [B_{\beta\alpha}]&[B_{\beta\beta}]\end{array}\right]=\left[\begin{array}{c|c}[C_{\alpha\alpha}]&[C_{\alpha\beta}]\\\hline [C_{\beta\alpha}]&[C_{\beta\beta}]\end{array}\right] \tag{1.42}$$

where

$$\begin{aligned}[C_{\alpha\alpha}]&=[A_{\alpha\alpha}][B_{\alpha\alpha}]+[A_{\alpha\beta}][B_{\beta\alpha}]\\ [C_{\alpha\beta}]&=[A_{\alpha\alpha}][B_{\alpha\beta}]+[A_{\alpha\beta}][B_{\beta\beta}]\\ [C_{\beta\alpha}]&=[A_{\beta\alpha}][B_{\alpha\alpha}]+[A_{\beta\beta}][B_{\beta\alpha}]\\ [C_{\beta\beta}]&=[A_{\beta\alpha}][B_{\alpha\beta}]+[A_{\beta\beta}][B_{\beta\beta}]\end{aligned} \tag{1.42a}$$

3. The inverse of a partitioned matrix $[A]$, can be obtained in terms of its submatrices. If $[C]$ is the inverse of $[A]$, i.e.,

$$[A]^{-1}=[C]$$

then in partitioned form

$$\left[\begin{array}{c|c}[A_{\alpha\alpha}]&[A_{\alpha\beta}]\\\hline [A_{\beta\alpha}]&[A_{\beta\beta}]\end{array}\right]^{-1}=\left[\begin{array}{c|c}[C_{\alpha\alpha}]&[C_{\alpha\beta}]\\\hline [C_{\beta\alpha}]&[C_{\beta\beta}]\end{array}\right] \tag{1.43}$$

Where the submatrices C_{ij} $(i, j=\alpha, \beta)$ are defined as

$$\begin{aligned}[C_{\alpha\alpha}]&=[[A_{\alpha\alpha}]-[A_{\alpha\beta}][A_{\beta\beta}]^{-1}[A_{\beta\alpha}]]^{-1}\\ [C_{\alpha\beta}]&=-[A_{\alpha\alpha}]^{-1}[A_{\alpha\beta}][C_{\beta\beta}]\\ [C_{\beta\alpha}]&=-[A_{\beta\beta}]^{-1}[A_{\beta\alpha}][C_{\alpha\alpha}]\\ [C_{\beta\beta}]&=[[A_{\beta\beta}]-[A_{\beta\alpha}][A_{\alpha\alpha}]^{-1}[A_{\alpha\beta}]]^{-1}\end{aligned} \tag{1.43a}$$

Example

It is required to find the inverse of the following partitioned matrix

$$[A] = \left[\begin{array}{cc|cc} 1 & 2 & 1 & 2 \\ 2 & 3 & 0 & 1 \\ \hline 1 & 0 & 4 & 2 \\ 2 & 1 & 2 & 2 \end{array}\right] \quad \begin{matrix} [A_{\alpha\alpha}] & [A_{\alpha\beta}] \\ [A_{\beta\alpha}] & [A_{\beta\beta}] \end{matrix}$$

From Eq. (1.27a)

$$[C_{\alpha\alpha}] = \left[\begin{bmatrix} 1 & 2 \\ 2 & 3 \end{bmatrix} - \begin{bmatrix} 1 & 2 \\ 0 & 1 \end{bmatrix}\begin{bmatrix} 4 & 2 \\ 2 & 2 \end{bmatrix}^{-1}\begin{bmatrix} 1 & 0 \\ 2 & 1 \end{bmatrix}\right]^{-1} = \frac{1}{13}\begin{bmatrix} -8 & 2 \\ 2 & 6 \end{bmatrix}$$

$$[C_{\beta\beta}] = \left[\begin{bmatrix} 4 & 2 \\ 2 & 2 \end{bmatrix} - \begin{bmatrix} 1 & 0 \\ 2 & 1 \end{bmatrix}\begin{bmatrix} 1 & 2 \\ 2 & 3 \end{bmatrix}^{-1}\begin{bmatrix} 1 & 2 \\ 0 & 1 \end{bmatrix}\right]^{-1} = \frac{1}{13}\begin{bmatrix} 7 & -6 \\ -6 & 7 \end{bmatrix}$$

$$[C_{\alpha\beta}] = \frac{1}{13}\begin{bmatrix} 1 & 2 \\ 2 & 3 \end{bmatrix}^{-1}\begin{bmatrix} 1 & 2 \\ 0 & 1 \end{bmatrix}\begin{bmatrix} 7 & -6 \\ 6 & 7 \end{bmatrix} = \frac{1}{13}\begin{bmatrix} -3 & 10 \\ 4 & 9 \end{bmatrix}$$

$$[C_{\beta\alpha}] = -\frac{1}{13}\begin{bmatrix} 4 & 2 \\ 2 & 2 \end{bmatrix}^{-1}\begin{bmatrix} 1 & 0 \\ 2 & 1 \end{bmatrix}\begin{bmatrix} -8 & 2 \\ 2 & 6 \end{bmatrix} = \frac{1}{13}\begin{bmatrix} -3 & 4 \\ 10 & -9 \end{bmatrix}$$

Finally, the inverse of the matrix is

$$[A]^{-1} = \frac{1}{13}\begin{bmatrix} -8 & 2 & -3 & 10 \\ 2 & 6 & 4 & -9 \\ -3 & 4 & 7 & -6 \\ 10 & -9 & -6 & 7 \end{bmatrix}$$

1.4 SIMULTANEOUS LINEAR ALGEBRAIC EQUATIONS

The use of matrices provides a neat systematic way of solving a set of linear algebraic equations. Let us consider, for instance, the following set of linear algebraic simultaneous equations in the unknowns $\delta i (i = 1, 2, 3, \ldots, m)$

$$\begin{aligned} C_{11}\delta_1 + C_{12}\delta_2 + C_{13}\delta_3 + \cdots C_{1m}\delta_m &= F_1 \\ C_{21}\delta_1 + C_{22}\delta_2 + C_{23}\delta_3 + \cdots C_{2m}\delta_m &= F_2 \\ C_{31}\delta_1 + C_{32}\delta_2 + C_{33}\delta_3 + \cdots C_{3m}\delta_m &= F_3 \\ \cdots & \\ \cdots & \\ \cdots & \\ C_{m1}\delta_1 + C_{m2}\delta_2 + C_{m3}\delta_3 + \cdots C_{mm}\delta_m &= F_m \end{aligned} \tag{1.44}$$

or in shorthand summation form Eq. (1.44) can be written as,

$$F_k = \sum_{n=1}^{m} C_{kn}\delta_n \qquad (k = 1, 2, \ldots, m). \tag{1.44a}$$

Equation (1.44) can also be written in matrix form as

$$\begin{bmatrix} C_{11} & C_{12} & C_{13} \cdots C_{1m} \\ C_{21} & C_{22} & C_{23} \cdots C_{2m} \\ C_{31} & C_{32} & C_{33} \cdots C_{3m} \\ \cdots & & \\ \cdots & & \\ C_{m1} & C_{m2} & C_{m3} \cdots C_{mm} \end{bmatrix} \begin{bmatrix} \delta_1 \\ \delta_2 \\ \delta_3 \\ \\ \\ \delta_m \end{bmatrix} = \begin{bmatrix} F_1 \\ F_2 \\ F_3 \\ \\ \\ F_m \end{bmatrix} \tag{1.45}$$

or in shorthand matrix notation,

$$[C]\{\delta\} = \{F\}. \tag{1.45a}$$

Upon premultiplying Eq. (1.45a) by the inverse of $[C]$, the following is obtained

$$[C]^{-1}[C]\{\delta\} = [C]^{-1}\{F\},$$

or

$$\{\delta\} = [C]^{-1}\{F\}.$$

Having the inverse, $[C]^{-1}$, the unknowns $\delta_1\delta_2, \ldots, \delta_m$ can be obtained.

Example

Suppose we have the following two linear equations,

$$\delta_1 + 2\delta_2 = 5,$$

$$3\delta_1 + 4\delta_2 = 11.$$

In matrix form this can be written as

$$\begin{bmatrix} 1 & 2 \\ 3 & 4 \end{bmatrix} \begin{bmatrix} \delta_1 \\ \delta_2 \end{bmatrix} = \begin{bmatrix} 5 \\ 11 \end{bmatrix}$$

or

$$\begin{bmatrix} \delta_1 \\ \delta_2 \end{bmatrix} = \begin{bmatrix} 1 & 2 \\ 3 & 4 \end{bmatrix}^{-1} \begin{bmatrix} 5 \\ 11 \end{bmatrix} = \begin{bmatrix} -2 & 1 \\ \frac{3}{2} & -\frac{1}{2} \end{bmatrix} \begin{bmatrix} 5 \\ 11 \end{bmatrix}$$

Expanding yields

$$\delta_1 = -2 \times 5 + 11 = 1$$

$$\delta_2 = \tfrac{3}{2}(5) - \tfrac{1}{2}(11) = 2$$

1.5 SUPPLEMENTARY PROBLEMS

Problem 1.1. Find the determinant of each of the following matrices

a. $\begin{bmatrix} 1 & 2 & 3 \\ 5 & 6 & 2 \\ 4 & 9 & 7 \end{bmatrix}$ b. $\begin{bmatrix} 2 & 3 & 5 \\ 8 & 9 & 1 \\ 4 & 0 & 2 \end{bmatrix}$ c. $\begin{bmatrix} A_{11} & A_{12} \\ & \\ A_{21} & A_{22} \end{bmatrix}$

Problem 1.2. Given the following matrices

$$[A]=\begin{bmatrix}1&2&3\\4&5&6\\7&8&9\end{bmatrix},\qquad [B]=\begin{bmatrix}3&1&4\\1&6&5\\6&5&8\end{bmatrix}.$$

Find:

a. $[S]=[A]+[B]$ (matrix summation)

b. $[D]=[A]-[B]$ (matrix subtraction)

c. $[P]=[A][B]$ (matrix multiplication)

d. $[M]=[B][A]$ (matrix multiplication)

Problem 1.3. Partition the matrices, $[A]$ and $[B]$, given in Problem 1.2 and find the sum, difference and product in terms of the submatrices.

Problem 1.4. Find the elements of matrix $[C]$ in terms of the submatrices as shown below

$$[C]=\begin{bmatrix}[C_{11}] & [C_{12}]\\ [C_{21}] & [C_{22}]\end{bmatrix}=\overset{[A]}{\left[\begin{array}{cc|c}A_{11} & A_{12} & A_{13}\\ A_{21} & A_{22} & A_{23}\\ \hline A_{31} & A_{32} & A_{33}\end{array}\right]}\times\overset{[B]}{\left[\begin{array}{c|c}B_{11} & B_{12}\\ B_{21} & B_{22}\\ \hline B_{31} & B_{32}\end{array}\right]}$$

Problem 1.5. In Problem 1.4 let

$$A_{11}=2,\quad A_{22}=4,\quad A_{33}=6$$

$$A_{12}=A_{21}=3,\quad A_{13}=A_{31}=8$$

$$A_{23}=A_{32}=1$$

$$B_{11}=2,\quad B_{22}=3$$

$$B_{12}=B_{21}=1$$

$$B_{31}=4,\quad B_{32}=6$$

a. Find $[C]$ by evaluating $[C_{11}]$, $[C_{12}]$, $[C_{21}]$, and $[C_{22}]$.

b. Remove the partitioning from $[A]$ and $[B]$ and use the numerical values for A_{ij} and B_{ij} to find $[C]$. Compare with (a).

Problem 1.6. Find the inverse for each of the following matrices.

a. $\begin{bmatrix}1&2&3\\2&4&5\\3&5&6\end{bmatrix}$ b. $\begin{bmatrix}1&2&3\\2&7&5\\2&4&6\end{bmatrix}$ c. $\begin{bmatrix}1&2&3\\2&4&5\\1&2&2\end{bmatrix}$

Problem 1.7. Given three linear simultaneous algebraic equations as

$$\delta_x+2\delta_y+5\delta_z=5$$

$$5\delta_x+7\delta_y+\delta_z=9$$

$$3\delta_x+\delta_y+4\delta_z=1$$

Form the matrix equation and solve for δ_x, δ_y and δ_z, i.e.,

$$[A]\{\delta\}=\{C\}$$

or

$$\{\delta\} = [A]^{-1}\{C\}$$

Problem 1.8. Verify Eq. (1.27a).

Problem 1.9. Given the matrix shown below, find its inverse in such a manner that you need not invert more than a 2×2 matrix.

$$\begin{bmatrix} 1 & 2 & 0 & 3 \\ 2 & 5 & 6 & 2 \\ 0 & 6 & 1 & 0 \\ 3 & 2 & 0 & 8 \end{bmatrix}$$

Problem 1.10. Use the Cholesky Method to find the inverse of nonsymmetric, nonsingular square matrix $[A]$.

Problem 1.11. Given the following partitioned matrix equation where the matrix $[A]$ is assumed to be known

$$\begin{bmatrix} F_\alpha \\ \hline F_\beta \end{bmatrix} = \left[\begin{array}{c|c} [A_{\alpha\alpha}]^{n\times n} & [A_{\alpha\beta}]^{n\times m} \\ \hline [A_{\beta\alpha}]^{m\times n} & [A_{\beta\beta}]^{m\times m} \end{array}\right] \begin{bmatrix} \delta_\alpha \\ \hline \delta_\beta \end{bmatrix}$$

a. Assume that $\{F_\alpha\}$ and $\{\delta_\beta\}$ are the known vectors, find by matrix manipulation the unknown vectors $\{F_\beta\}$ and $\{\delta_\alpha\}$.
b. Knowing that $n = m = 2$, $\delta_\beta = 0$ and

$$[A_{\alpha\alpha}] = \begin{bmatrix} 3 & 1 \\ 2 & 1 \end{bmatrix}$$

Find the unknown elements of δ_α, i.e., δ_1 and δ_2. Use matrix technique and show step by step procedure.

1.6 SUGGESTED COMPUTER EXERCISES

1. Write a computer program to print-out the elements of a matrix $[S]$ whose maximum order is 100×100.

2. Write a computer program to perform the following matrix operations

a. $[A]^{m\times i} \times [B]^{i\times n} = [C]^{m\times n}$

b. $[A]^{m\times n} - [B]^{m\times n} = [D]^{m\times n}$

c. $[A]^{m\times n} + [B]^{m\times n} = [S]^{m\times n}$

3. Given a square matrix $[A]^{n\times n}$ whose inverse is denoted by $[C]^{n\times n}$, write a computer program to find the inverted matrix $[C]$.

4. Given a set of 100 linear simultaneous equations in the unknowns $q_n (n = 1, 2, \ldots, 100)$,

$$[S]^{n\times n}\{q\}^{n\times 1} = \{F\}^{n\times 1},$$

write a computer program to evaluate the unknowns $q_n (n = 1, 2, \ldots, 100)$ assuming that $[S]$ and $\{F\}$ are given.

CHAPTER TWO

Fundamental Concepts and Principles

2.1 INTRODUCTION

The objective of this chapter is to introduce the fundamental structural concepts and principles which are directly related to the development of the matrix methods in subsequent chapters. The concepts of generalized forces and displacements, structural behavior, constituent structural elements, strain energy, Castigliano's Theorems, Maxwell Reciprocal Theorem, methods of superposition, etc., are introduced. It is important that the reader understand the material presented herein.

2.2 GENERALIZED FORCES AND DISPLACEMENTS

The generalized force, Q, is used in this text to represent forces, moments and/or torques. For instance, the force vector for the beam element (Fig. 2.3a) is represented by

$$\{Q\} = \begin{bmatrix} Q_1 \\ Q_2 \\ Q_3 \\ Q_4 \\ Q_5 \\ Q_6 \\ \vdots \\ Q_{12} \end{bmatrix} = \begin{bmatrix} F_i^x \\ F_i^y \\ F_i^z \\ M_i^x \\ M_i^y \\ M_i^z \\ \vdots \\ M_j^z \end{bmatrix} \tag{2.1}$$

Likewise, the generalized displacement, q, is used to represent translational and rotational displacements. Thus, the coordinate displacement vector for the beam shown in Fig. 2.3a may be expressed as follows:

$$\{q\} = \begin{bmatrix} q_1 \\ q_2 \\ q_3 \\ q_4 \\ q_5 \\ q_6 \\ \vdots \\ q_{12} \end{bmatrix} = \begin{bmatrix} \delta_i^x \\ \delta_i^y \\ \delta_i^z \\ \theta_i^x \\ \theta_i^y \\ \theta_i^z \\ \vdots \\ \theta_j^z \end{bmatrix} \tag{2.2}$$

2.3 STRUCTURAL BEHAVIOR

In accordance to the functional relationship between an applied force and the resulting displacement, structural behavior may be divided into three different classes:

a. Inelastic nonlinear.
b. Elastic nonlinear.
c. Linear elastic.

Consider a steadily loaded spring as shown below.

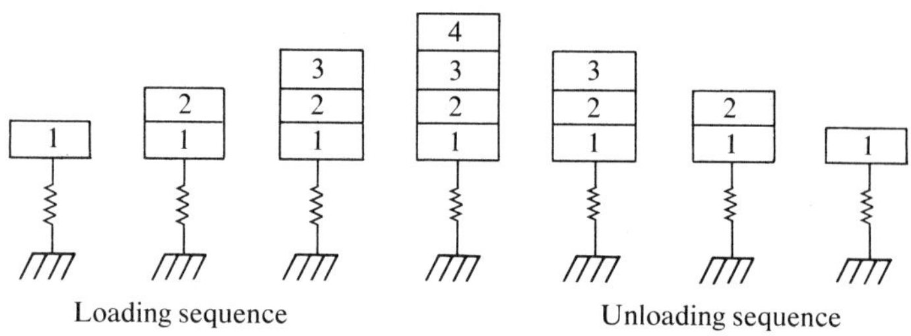

Fig. 2.1. Loading and unloading a spring.

If upon loading, the functional relationship between the force and its corresponding displacement takes on a curved path, and upon unloading it takes a different curved path as in Fig. 2.2(a), then the structural behavior is referred to as *inelastic nonlinear* behavior. If such functional relationship follows the same curved path in both loading and unloading conditions as in Fig. 2.2(b), then the behavior is said to be *elastic nonlinear* behavior.

On the other hand, if the functional relationship between the applied force and the resulting corresponding displacement takes on the same straight path as in Fig. 2.2(c), then such behavior is termed *linear elastic* behavior.

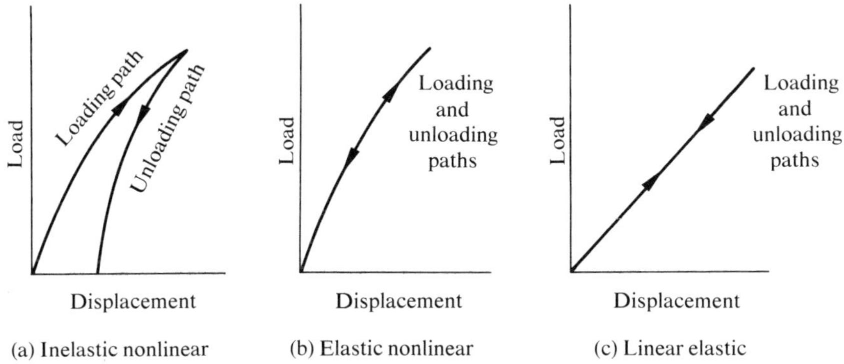

Fig. 2.2. Structural behavior.

The matrix techniques for structural analysis as presented in this text are to be used only in conjunction with structures whose behavior can be assumed linearly elastic unless otherwise specified. It is important that this assumption be adhered to, and careful measures must be taken if it is violated.

2.4 STRUCTURAL SYSTEM AND ITS CONSTITUENT ELEMENTS

Any system which has the ability of carrying and transmitting loads is characterized as a structural system or commonly referred to as structure. The loads may be static, dynamic, thermal loads or a combination of static, dynamic, and thermal loads.

The basic constituent elements of a structural system are beams, plates, shells or a combination of beams, plates, and shells.

Beam elements are in general structural members which can resist axial loads, shear loads, bending loads, and/or torsional loads. Figure 2.3(a) shows a typical beam element. Beams which are designed to resist axial loads only, such as in pinned-type trusses, are referred to in this text as *axial rods*. On the other hand, beams which transmit torsional loads only, as in a motor shaft, will be referred to as *torsional rods*.

Plate elements are two-dimensional generalization of beam elements. Plates which carry in-plane shearing forces only will be called *shear-panels*; if the loads are limited to in-plane normal forces then the plate will be referred to as a *membrane panel*.

Shell elements, on the other hand, may be defined as curved surfaces which enclose a space. Shells such as balloons, which can carry membrane type forces are referred to as membrane shells. (*See* Fig. 2.3 for illustration.)

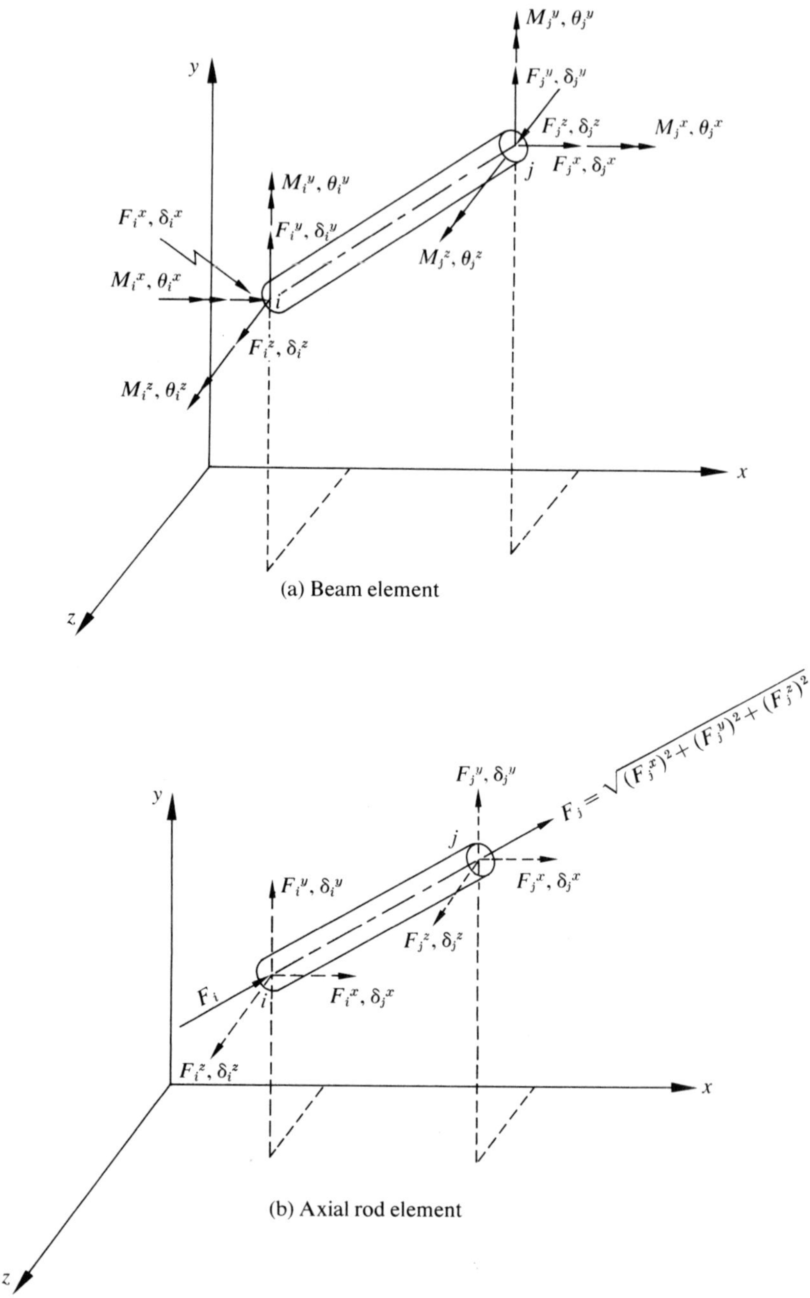

(a) Beam element

(b) Axial rod element

Fig. 2.3. Structural elements.

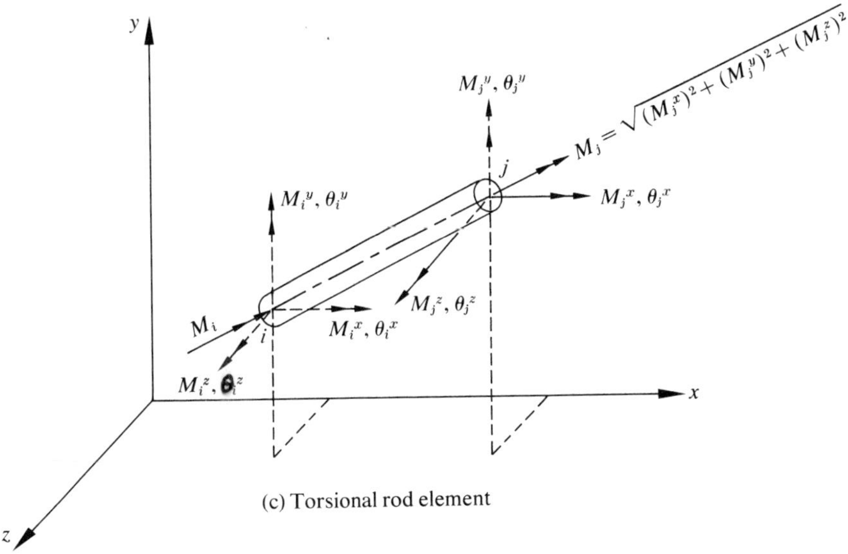

(c) Torsional rod element

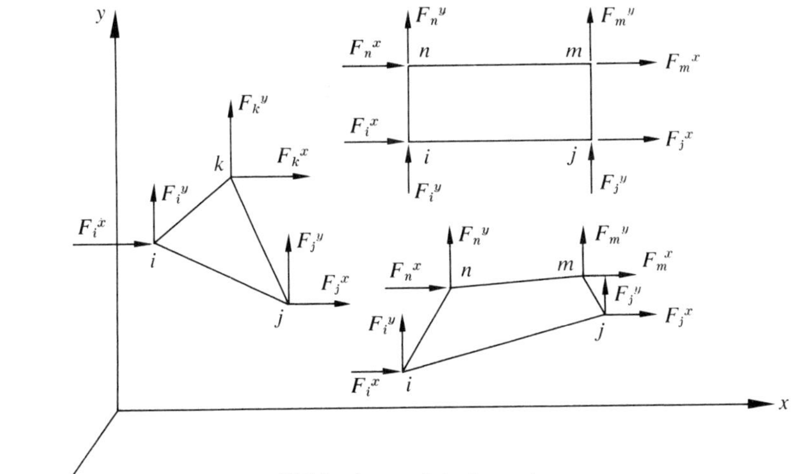

(d) Membrane plate elements

(2-D STRUCTURE)

Fig. 2.3. Structural elements.

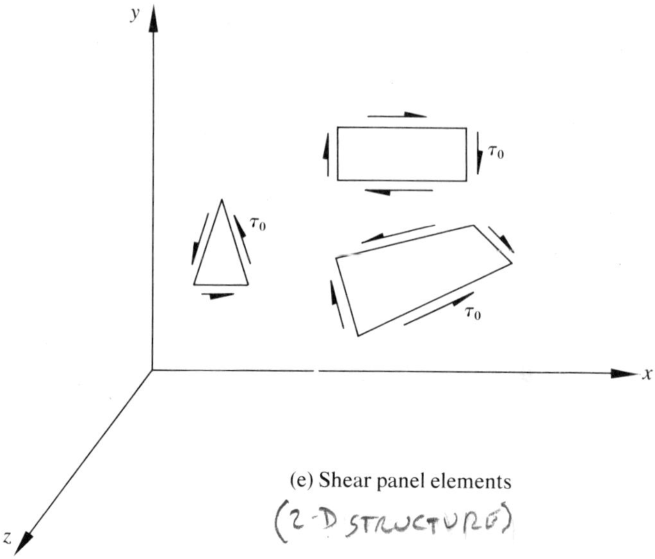

(e) Shear panel elements

(2-D STRUCTURE)

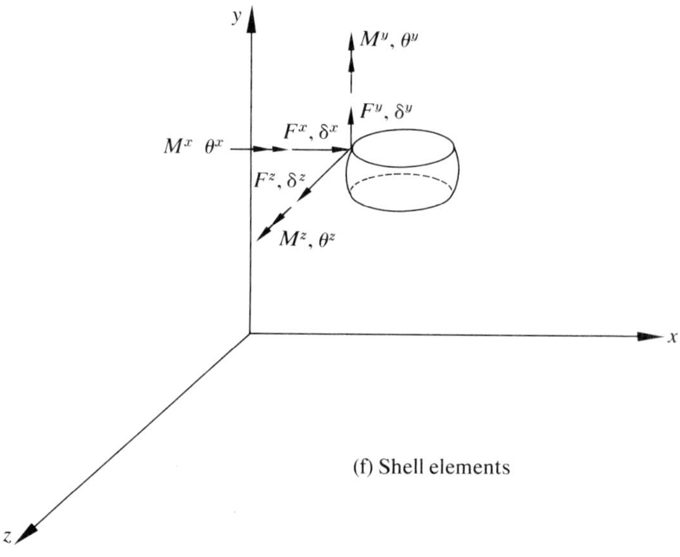

(f) Shell elements

Fig. 2.3. Structural elements.

2.5 ENERGY THEOREMS IN STRUCTURAL ANALYSIS

Exact elasticity solutions for complex structures are often difficult to obtain, if not impossible. Discontinuities, such as cutouts, abrupt change in thicknesses, different materials, etc., contribute to this fact. As an alternative, the structural analyst has reverted to the utilization of various energy theorems in obtaining practical solutions for many complex structural problems. The principles of virtual displacements, and virtual forces form the basis for all the energy theorems. In this text we will be concerned only with Castigliano's Theorems.

Consider the force-displacement curve as shown in Fig. 2.4, which for generality purposes is taken to be elastically nonlinear. The area under the curve (shaded vertically) is the work done on the system by the external force Q_i through a displacement q_i. If the system is under static equilibrium conditions, and if heat losses are neglected, then we can state that the work done is equal to the strain energy stored in the system, i.e.,

$$W = U \tag{2.3}$$

in which W = work and U = strain energy.

The area above the curve (shaded horizontally) is referred to as the complementary work and is equal to the complementary strain energy, i.e.,

$$\bar{W} = \bar{U}, \tag{2.4}$$

where $\bar{W}$ = complementary work and $\bar{U}$ = complementary strain energy.

If the system is given a virtual displacement, δq_i, then the virtual work δW, done by the force Q_i in moving through δq_i is,

$$\delta W = Q_i \delta q_i, \tag{2.5}$$

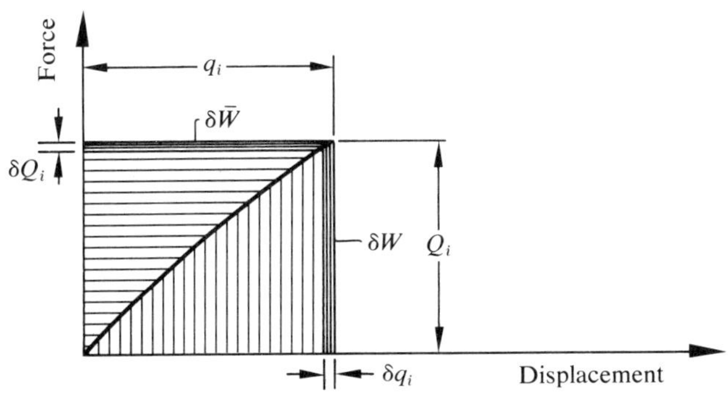

Fig. 2.4. Work and complementary work.

and from Eq. (2.3), it immediately follows that

$$\delta U = Q_i \delta q_i, \tag{2.6}$$

which in the limit becomes

$$\frac{\partial U}{\partial q_i} = Q_i. \tag{2.7}$$

Equation (2.7) is the well known first Castigliano's Theorem. In fact, it can be stated that the partial derivative of the strain energy with respect to any displacement q_i gives the force Q_i at point i in the direction of q_i.

On the other hand, if the system is subjected to a virtual force δQ_i in the direction of the displacement q_i, then the complementary virtual work is

$$\delta \bar{W} = q_i \delta Q_i. \tag{2.8}$$

From Eq. (2.4) it follows that

$$\delta \bar{U} = q_i \delta Q_i, \tag{2.9}$$

which in the limit becomes

$$\frac{\partial \bar{U}}{\partial Q_i} = q_i. \tag{2.10}$$

Equation (2.10) represents the widely known second theorem of Castigliano. It states that the partial derivative of the complementary strain energy with respect to any force Q_i at a point i, gives the displacement q_i at i in the direction of Q_i.

Assuming no heat dissipation in the system, an important relationship for linear-elastic structures, is that the strain and complementary strain energies are equal, i.e.,

$$U = \bar{U}. \tag{2.11}$$

Utilizing Eq. (2.11) into Eq. (2.10), yields

$$\frac{\partial U}{\partial Q_i} = q_i. \tag{2.12}$$

It is essential to remember that Eq. (2.12) is valid only to the extent of the above stated assumptions, i.e., the structure must be linearly elastic and temperature change effects can be neglected.

2.6 BASIC THERMO-ELASTIC RELATIONS

Stress-Strain Relations

The total strain in an elastic body under the action of both external and thermal loading can be expressed as follows

$$\epsilon_{\alpha\beta} = \bar{\epsilon}_{\alpha\beta} + \delta_{\alpha\beta}\epsilon^t_{\alpha\beta} \qquad (\alpha, \beta = x, y, z) \tag{2.13}$$

where $\delta_{\alpha\beta} = \begin{cases} 1 & \alpha = \beta \\ 0 & \alpha \neq \beta \end{cases}$;

$\epsilon_{\alpha\beta}$ = total strain;
$\bar{\epsilon}_{\alpha\beta}$ = elastic strain due to external applied loads;
$\epsilon^t_{\alpha\beta}$ = thermal strains due to thermal loads.

For an isotropic construction, the elastic strains are given by Hooke's Law as

$$\begin{aligned}
\bar{\epsilon}_{xx} &= \frac{1}{E}[\sigma_{xx} - \nu(\sigma_{yy} + \sigma_{zz})] \\
\bar{\epsilon}_{yy} &= \frac{1}{E}[\sigma_{yy} - \nu(\sigma_{xx} + \sigma_{zz})] \\
\bar{\epsilon}_{zz} &= \frac{1}{E}[\sigma_{zz} - \nu(\sigma_{zz} + \sigma_{yy})] \\
\bar{\epsilon}_{xy} &= \frac{1}{G}[\sigma_{xy}] = \frac{2(1+\nu)}{E}\sigma_{xy} \\
\bar{\epsilon}_{yz} &= \frac{1}{G}[\sigma_{yz}] = \frac{2(1+\nu)}{E}\sigma_{yz} \\
\bar{\epsilon}_{zx} &= \frac{1}{G}[\sigma_{zx}] = \frac{2(1+\nu)}{E}\sigma_{zx}
\end{aligned} \tag{2.14}$$

The thermal strains are defined as

$$\epsilon^t_{\alpha\beta} = \mu T \qquad (\alpha, \beta = x, y, z) \tag{2.15}$$

where E = Young's Modulus;
ν = Poisson's Ratio;
μ = coefficient of thermal expansion of material;
T = temperature change above a given datum temperature.

Thus, utilizing Eqs. (2.14) and (2.15) in Eq. (2.13) yields the three-dimensional thermo-elastic strain-stress relations

$$\begin{aligned}
\epsilon_{xx} &= \frac{1}{E}[\sigma_{xx} - \nu(\sigma_{yy} + \sigma_{zz})] + \mu T \\
\epsilon_{yy} &= \frac{1}{E}[\sigma_{yy} - \nu(\sigma_{xx} + \sigma_{zz})] + \mu T \\
\epsilon_{zz} &= \frac{1}{E}[\sigma_{zz} - \nu(\sigma_{xx} + \sigma_{yy})] + \mu T \\
\epsilon_{xy} &= \frac{2(1+\nu)}{E}(\sigma_{xy}) \\
\epsilon_{yz} &= \frac{2(1+\nu)}{E}(\sigma_{yz}) \\
\epsilon_{zx} &= \frac{2(1+\nu)}{E}(\sigma_{zx})
\end{aligned} \tag{2.16}$$

Equations (2.16) can be written in matrix notation as

$$\begin{bmatrix}\epsilon_{xx}\\ \epsilon_{yy}\\ \epsilon_{zz}\\ \epsilon_{xy}\\ \epsilon_{yz}\\ \epsilon_{zx}\end{bmatrix} = \frac{1}{E}\begin{bmatrix}1 & & & & & \\ -\nu & 1 & & \text{Sym.} & & \\ -\nu & -\nu & 1 & & & \\ 0 & 0 & 0 & 2(1+\nu) & & \\ 0 & 0 & 0 & 0 & 2(1+\nu) & \\ 0 & 0 & 0 & 0 & 0 & 2(1+\nu)\end{bmatrix} \times \begin{bmatrix}\sigma_{xx}\\ \sigma_{yy}\\ \sigma_{zz}\\ \sigma_{xy}\\ \sigma_{yz}\\ \sigma_{zx}\end{bmatrix} + \mu T\begin{bmatrix}1\\1\\1\\0\\0\\0\end{bmatrix} \tag{2.17}$$

Or in matrix compact form as

$$\{\mathrm{E}\} = [\mathrm{N}]\{\Sigma\} + \{\mathrm{M}\} \tag{2.17a}$$

Premultiplying Eq. (2.17a) by the inverse of N, will yield expressions for the stresses in terms of total strains as

$$\begin{bmatrix}\sigma_{xx}\\ \sigma_{yy}\\ \sigma_{zz}\\ \sigma_{xy}\\ \sigma_{yz}\\ \sigma_{zx}\end{bmatrix} = \lambda\begin{bmatrix}\eta_1 & & & & & \\ \nu & \eta_1 & & & & \\ \nu & \nu & \eta_1 & & & \\ 0 & 0 & 0 & \frac{1}{2}\eta_2 & & \\ 0 & 0 & 0 & 0 & \frac{1}{2}\eta_2 & \\ 0 & 0 & 0 & 0 & 0 & \frac{1}{2}\eta_2\end{bmatrix}\begin{bmatrix}\epsilon_{xx}\\ \epsilon_{yy}\\ \epsilon_{zz}\\ \epsilon_{xy}\\ \epsilon_{yz}\\ \epsilon_{zx}\end{bmatrix} - \frac{\mu E T}{\eta_2}\begin{bmatrix}1\\1\\1\\0\\0\\0\end{bmatrix} \tag{2.18}$$

where

$$\lambda = E/(1+\nu)(1-2\nu)$$

and

$$\eta_1 = 1-\nu; \qquad \eta_2 = 1-2\nu$$

Equation (2.18) can be written in compact matrix form as

$$\{\Sigma\} = [\Theta]\{\mathrm{E}\} - [\Theta]\{\mathrm{M}\} \tag{2.18a}$$

in which

$$\{\Sigma\};\{\mathrm{E}\};\{\mathrm{M}\} = \begin{bmatrix}\sigma_{xx}\\ \sigma_{yy}\\ \sigma_{zz}\\ \sigma_{xy}\\ \sigma_{yz}\\ \sigma_{zx}\end{bmatrix};\begin{bmatrix}\epsilon_{xx}\\ \epsilon_{yy}\\ \epsilon_{zz}\\ \epsilon_{xy}\\ \epsilon_{yz}\\ \epsilon_{zx}\end{bmatrix};\mu T\begin{bmatrix}1\\1\\1\\0\\0\\0\end{bmatrix} \tag{2.19}$$

and

$$[\Theta] = \lambda\begin{bmatrix}\eta_1 & & & & & \\ \nu & \eta_1 & & \text{Sym.} & & \\ \nu & 0 & \eta_1 & & & \\ 0 & 0 & 0 & \frac{1}{2}\eta_2 & & \\ 0 & 0 & 0 & 0 & \frac{1}{2}\eta_2 & \\ 0 & 0 & 0 & 0 & 0 & \frac{1}{2}\eta_2\end{bmatrix} \tag{2.20}$$

If the two-dimensional plane stress distribution prevails, (i.e. $\sigma_{zz} = \sigma_{zx} = \sigma_{yz} = 0$), then Eq. (2.18) reduces to

$$\begin{bmatrix}\sigma_{xx}\\ \sigma_{yy}\\ \sigma_{xy}\end{bmatrix} = \bar{E}\begin{bmatrix}1 & & \text{Sym.}\\ \nu & 1 & \\ 0 & 0 & \frac{1}{2}\eta_1\end{bmatrix}\begin{bmatrix}\epsilon_{xx}\\ \epsilon_{yy}\\ \epsilon_{xy}\end{bmatrix} - \frac{\mu E T}{\eta_1}\begin{bmatrix}1\\ 1\\ 0\end{bmatrix}, \qquad \bar{E} = E/(1-\nu^2) \tag{2.21}$$

Solving Eq. (2.21) for the strains yields

$$\begin{bmatrix}\epsilon_{xx}\\ \epsilon_{yy}\\ \epsilon_{xy}\end{bmatrix} = \frac{1}{E}\begin{bmatrix}1 & & \text{Sym.}\\ -\nu & 1 & \\ 0 & 0 & 2(1+\nu)\end{bmatrix}\begin{bmatrix}\sigma_{xx}\\ \sigma_{yy}\\ \sigma_{xy}\end{bmatrix} + \mu T\begin{bmatrix}1\\ 1\\ 0\end{bmatrix} \tag{2.22}$$

It is evident from Eq. (2.16) that ϵ_{zz} for plane stress distribution becomes

$$\epsilon_{zz} = -\frac{\nu}{E}(\sigma_{xx} + \sigma_{yy}) + \mu T \tag{2.23}$$

In the case of two-dimensional plane strain distribution (i.e., $\epsilon_{zz} = \epsilon_{yz} = \epsilon_{zx} = 0$) Eq. (2.18) reduces to

$$\begin{bmatrix}\sigma_{xx}\\ \sigma_{yy}\\ \sigma_{xy}\end{bmatrix} = \lambda\begin{bmatrix}\eta_1 & & \text{Sym.}\\ \nu & \eta_1 & \\ 0 & 0 & \frac{1}{2}\eta_2\end{bmatrix}\begin{bmatrix}\epsilon_{xx}\\ \epsilon_{yy}\\ \epsilon_{xy}\end{bmatrix} - \frac{\mu E T}{\eta_2}\begin{bmatrix}1\\ 1\\ 0\end{bmatrix} \tag{2.24}$$

and

$$\sigma_{zz} = \lambda\nu(\epsilon_{xx} + \epsilon_{yy}) - \frac{\mu E T}{\eta_2}. \tag{2.25}$$

For one-dimensional state of stress distribution such as in one-dimensional beam elements, the only non-zero stress can be expressed as

$$\sigma_{xx} = E\epsilon_{xx} - E\mu T, \tag{2.26}$$

or

$$\epsilon_{xx} = \frac{\sigma_{xx}}{E} + \mu T = \bar{\epsilon}_{xx} + \mu T. \tag{2.27}$$

Strain-Displacement Relations

If we let q_x, q_y and q_z be the generalized displacements in the x, y, and z directions, respectively, then the strains may be written as follows

$$\begin{aligned}&\epsilon_{xx} = q_{x,x}; \qquad \epsilon_{yy} = q_{y,y}; \qquad \epsilon_{zz} = q_{z,z};\\ &\epsilon_{xy} = q_{x,y} + q_{y,x}; \qquad \epsilon_{zx} = q_{x,z} + q_{z,x};\end{aligned} \tag{2.28}$$

and

$$\epsilon_{yz} = q_{y,z} + q_{z,y}.$$

The comma here indicates partial differentiation with respect to the following subscript.

Equations (2.28) are the three-dimensional linearized strain displacement relations and are based on the assumption of small deformation.

2.7 STRAIN AND COMPLEMENTARY STRAIN ENERGIES

The strain and complementary strain energies for linear-elastic structures can be expressed in terms of stresses and strain as,

$$U = \bar{U} = \tfrac{1}{2} \int_v \lfloor \Sigma \rfloor \{\mathrm{E}\} dV = \tfrac{1}{2} \int_v \lfloor \mathrm{E} \rfloor \{\Sigma\} dV \tag{2.29}$$

in which

$$\{\mathrm{E}\}^T = \lfloor \mathrm{E} \rfloor = \lfloor \epsilon_{xx}\, \epsilon_{yy}\, \epsilon_{zz}\, \epsilon_{xy}\, \epsilon_{yz}\, \epsilon_{zx} \rfloor$$

$$\{\Sigma\}^T = \lfloor \Sigma \rfloor = \lfloor \sigma_{xx}\, \sigma_{yy}\, \sigma_{zz}\, \sigma_{xy}\, \sigma_{yz}\, \sigma_{zx} \rfloor$$

and $\int_v dV$ implies integration over the total volume of the structural element under consideration.

Assuming that temperature change is zero and utilizing Eq. (2.17a) the expression for U and $\bar{U}$ can be put in terms of stresses alone

$$U = \bar{U} = \tfrac{1}{2} \int_v \lfloor \Sigma \rfloor [\mathrm{N}] \{\Sigma\} dV \tag{2.30}$$

or in terms of strains by utilizing Eq. (2.18a)

$$U = \bar{U} = \tfrac{1}{2} \int_v \lfloor \mathrm{E} \rfloor [\Theta] \{\mathrm{E}\} dV \tag{2.31}$$

The strain energy due to elastic strains for a beam element can be expressed in terms of internal forces as,

$$U = \bar{U} = \tfrac{1}{2} \int_{s_1}^{s_2} \left\{ \frac{1}{EI(s)} [M(s)]^2 + \frac{1}{EA(s)} [S(s)]^2 + \frac{1}{GJ(s)} [T(s)]^2 + \frac{1}{GA(s)} [V(s)]^2 \right\} ds \tag{2.32}$$

where s = coordinate axis taken along the neutral axis of the beam; E and G = Young's and shear moduli, respectively; $I(s)$, $A(s)$, $J(s)$ = area moment of inertia, area and torsional moment of inertia of the beam, respectively; $M(s)$, $S(s)$, $T(s)$ and $V(s)$ = bending moment, axial force, torsional moment and transverse shear force at a distance s along the beam, respectively,

$$\tfrac{1}{2} \int \frac{1}{EI(s)} [M(s)]^2 ds = \text{strain energy due to bending,}$$

$$\tfrac{1}{2} \int \frac{1}{EA(s)} [S(s)]^2 ds = \text{strain energy due to axial loading,}$$

$$\tfrac{1}{2}\int \frac{1}{GJ(s)}[T(s)]^2 ds = \text{strain energy due to torsion,}$$

and

$$\tfrac{1}{2}\int \frac{1}{GA(s)}[V(s)]^2 ds = \text{strain energy due to transverse shear.}$$

In the case of a constant shear flow rectangular panel, the strain energy is expressed by,

$$U = \bar{U} = \tfrac{1}{2}\int\int \frac{1}{Gt}\tau_0{}^2 dxdy = \tfrac{1}{2}\frac{L_x L_y \tau_0{}^2}{Gt} \qquad (2.33)$$

where L_x and L_y = length of panel in x- and y-directions, respectively;
τ_0 = constant shear flow per unit length;
t = thickness of panel.

Illustrative Examples

Example 1

Find the strain energy stored in the bar shown below. Assume the load is acting along the neutral axis of the bar. From Eq. (2.32) the strain energy due to axial loading is,

$$U = \tfrac{1}{2}\int \frac{S^2 dx}{AE}$$

where $S = P =$ constant.

$$A = A_0 e^{\alpha x}$$

Hence

$$U = \tfrac{1}{2}\int_0^L \frac{p^2}{A_0 e^{\alpha x} E}dx = \frac{p^2}{2A_0 E}(1 - e^{-L\alpha})$$

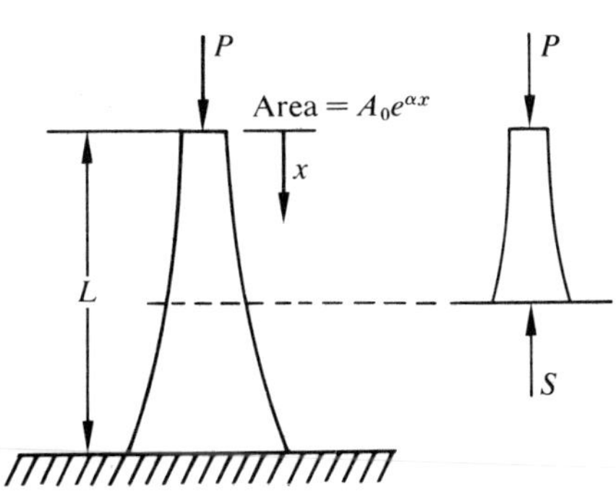

Example 2

Find the strain energy stored in the beam shown below. From Eq. (2.32) the strain energy due to bending and transverse shear is,

$$U = \tfrac{1}{2}\int_0^L \frac{[M(x)]^2dx}{EI_{ab}} + \tfrac{1}{2}\int_L^{2L} \frac{[M(x)]^2dx}{EI_{bc}} + \tfrac{1}{2}\int_0^L \frac{[V(x)]^2dx}{A_{ab}G} + \tfrac{1}{2}\int_L^{2L} \frac{[V(x)]^2dx}{A_{bc}G}.$$

Where $M(x) = Fx$
$V(x) = F$

$$U = \tfrac{1}{2}\int_0^L \frac{F^2x^2dx}{EI_0} + \tfrac{1}{2}\int_L^{2L} \frac{F^2x^2dx}{2EI_0} + \tfrac{1}{2}\int_0^L \frac{F^2dx}{AG} + \tfrac{1}{2}\int_L^{2L} \frac{F^2dx}{2AG}$$

$$= \frac{3F^2L^3}{4EI_0}\left(1 + \frac{EI_0}{L^2AG}\right)$$

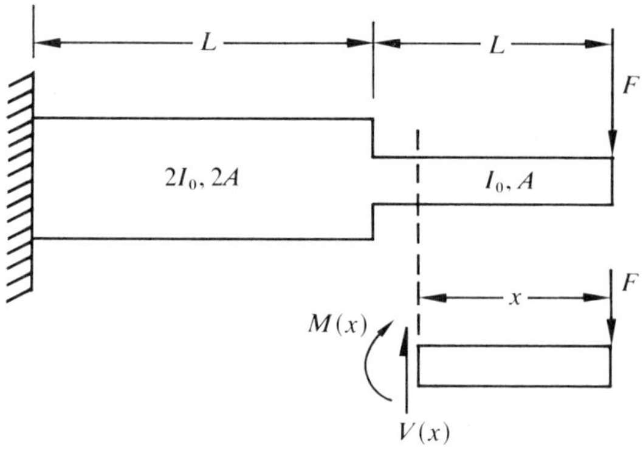

Example 3

Find the strain energy in the structure shown. In this particular example all four contributions to the strain energy are present. The internal forces are,

1. Section 1–2.

$$S(y) = F; \qquad M(y) = V(y) = T(y) = 0.$$

2. Section 2–3.

$$S(z) = 0; \qquad M(z) = Fz; \qquad V(z) = F; \qquad T(z) = 0.$$

3. Section 3–4.

$$S(x) = 0; \qquad M(x) = Fx; \qquad V(x) = F; \qquad T(x) = Fb.$$

Hence, substituting into Eq. (2.32) yields,

$$U=\tfrac{1}{2}\int_1^2 \frac{F^2dy}{AE}+\tfrac{1}{2}\int_2^3 \frac{F^2z^2dz}{EI}+\tfrac{1}{2}\int_2^3 \frac{F^2dz}{AG}+\tfrac{1}{2}\int_3^4 \frac{F^2x\,dx}{EI}$$

$$+\tfrac{1}{2}\int_3^4 \frac{F^2dx}{AG}+\tfrac{1}{2}\int_3^4 \frac{F^2b^2dx}{GJ}$$

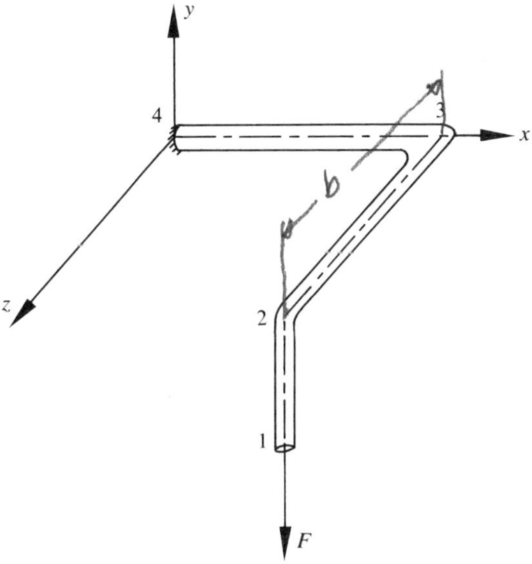

2.8 PRINCIPLES OF SUPERPOSITION

An important principle utilized in the solution of linearly-elastic structures is that of superposition of forces and displacements. The principle can best be explained by considering some illustrative examples.

Superposition of Forces

To illustrate the superposition of forces let us consider a cantilever beam subjected to a set of forces as shown in Fig. 2.5(a), where the displacement, q_i, is required.

If each force Q_n $(n=1, 2, \ldots, i)$ is applied to the beam one at a time, Fig. 2.4 $(b, c, \ldots, d)$ and the corresponding displacement is calculated at station i, then by superposition the total displacement at i can be expressed as,

$$q_i = q_i^{(1)} + q_i^{(2)} + \cdots + q_i^{(i)} \tag{2.34}$$

where $q_i^{(i)}$ = the displacement at i caused by a force Q_i.

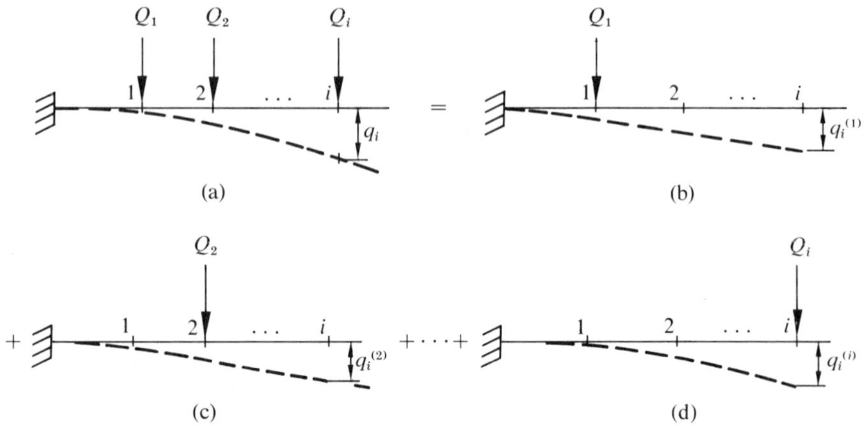

Fig. 2.5. Superposition of forces.

Equation (2.34) can also be written in terms of forces as

$$q_i = \alpha_{i1}Q_1 + \alpha_{i2}Q_2 + \cdots + \alpha_{ij}Q_j \tag{2.35}$$

or

$$q_i = \sum_{k=1}^{j} \alpha_{ik}Q_k \tag{2.35a}$$

in which α_{ik} are the influence coefficients and defined as the deflection at i caused by a unit force at k.

Superposition of Displacement

To illustrate the method of superposition of displacements let us take a beam as shown in Fig. 2.6(a) where the ends, 1 and 2, are given the displacements shown. The force, Q_i $(i = 1, 2, 3, 4)$, can be calculated by evaluating the forces caused by each displacement separately and then superimposing the results.

Thus in general

$$Q_i = Q_i^{(4)} + Q_i^{(2)} \qquad (i = 1, 2, 3, 4) \tag{2.36}$$

in which the bracketed superscripts in Eq. (2.36) denote the contribution of the displacements q_4 and q_2 to the forces at locations 1 and 2 caused by these displacements.

Equations (2.36) can be expressed in terms of displacements as

$$Q_i = S_{i1}q_4 + S_{i2}q_2 \qquad (i = 1, 2, 3, 4) \tag{2.37}$$

in which S_{i4} and S_{i2} are stiffness coefficients.

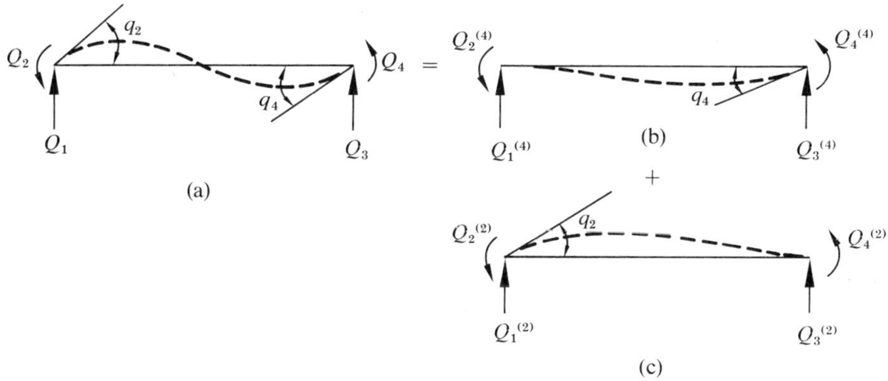

Fig. 2.6. Superposition of displacements.

In general a stiffness coefficient S_{ij} is defined as the force caused at location i due to a unit displacement at location j.

2.9 RECIPROCAL THEOREMS

a. Betti's Theorem.

Betti's Theorem is stated without proof as follows

In a linearly elastic structure the work done by a system of forces Q_I through the displacements q_{II} is equal to the work done by a system of forces Q_{II} through the displacements q_I, in which q_I and q_{II} are the displacements due to Q_I and Q_{II}, respectively.

b. Maxwell's Reciprocal Theorem.

If in Betti's Theorem we let the systems of forces Q_I and Q_{II} each consist of a single force, Q_i and Q_j, for instance, applied at locations i and j, respectively, (Fig. 2.7), then the work done (if Q_i is applied first then Q_j) is,

$$W_{i,j} = \tfrac{1}{2}\alpha_{ii}Q_i^2 + \alpha_{ij}Q_iQ_j + \tfrac{1}{2}\alpha_{jj}Q_j^2 \qquad (2.38a)$$

If Q_j is applied first then Q_i, the work done is,

$$W_{j,i} = \tfrac{1}{2}\alpha_{ii}Q_i^2 + \alpha_{ij}Q_iQ_j + \tfrac{1}{2}\alpha_{jj}Q_j^2 \qquad (2.38b)$$

in which α_{ii}, α_{ij} and α_{jj} are the influence coefficients.

From Betti's Theorem,

$$W_{i,j} = W_{j,i}$$

therefore we conclude that in general,

$$\alpha_{ij} = \alpha_{ji}. \qquad (2.39)$$

Likewise, it can be shown that,

$$S_{ij} = S_{ji}. \tag{2.40}$$

Equations (2.39) and (2.40) are referred to as the Maxwell Reciprocal Relations. Equation (2.39) states that the displacement at location i caused by a unit force at j is the same as the displacement at j caused by a unit force at i. On the other hand, Eq. (2.40) states that the force at i caused by a unit displacement at j is the same as the force at j caused by a unit displacement at i.

For illustration, consider the beam shown below.

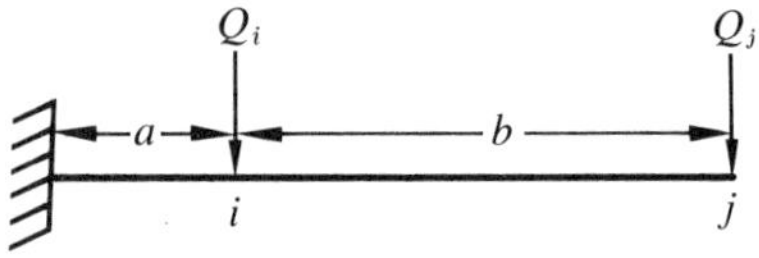

Fig. 2.7. Force system.

From elementary strength of materials, the deflection, q_j, at j, due to a unit force Q_i alone is given by

$$q_j = \frac{1}{6\,EI}(3a^2L - a^3). \tag{a}$$

Similarly, the deflection at i due to a unit force Q_j alone is,

$$q_i = \frac{1}{6\,EI}(b^3 - 3L^2b + 2L^3) = \frac{1}{6\,EI}(L-a)^3 - 3L^2(L-a) + 2L^3$$

$$= \frac{1}{6\,EI}(3a^2L - a^3) \tag{b}$$

Obviously the right-hand side of Eqs. (a) and (b) are equal, hence $q_i = q_j$ which illustrates Maxwell's Reciprocal Theorem.

2.10 SUPPLEMENTARY PROBLEMS

Problem 2.1. Use the method of superposition to find the vertical displacement at (b) and the slope at (a) of the structure shown. Assume EI to be constant.

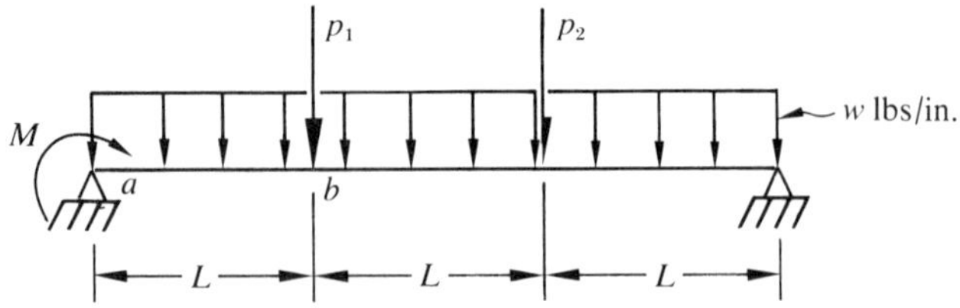

Problem 2.2. Use the method of superposition to find the unknown reactions of the structure shown.

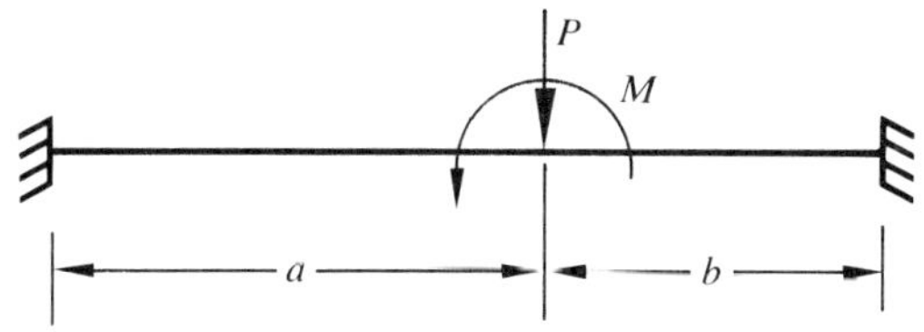

Problem 2.3. Use the method of superposition to find the resulting forces due to the deflected shape of the structure shown. Assume constant EI.

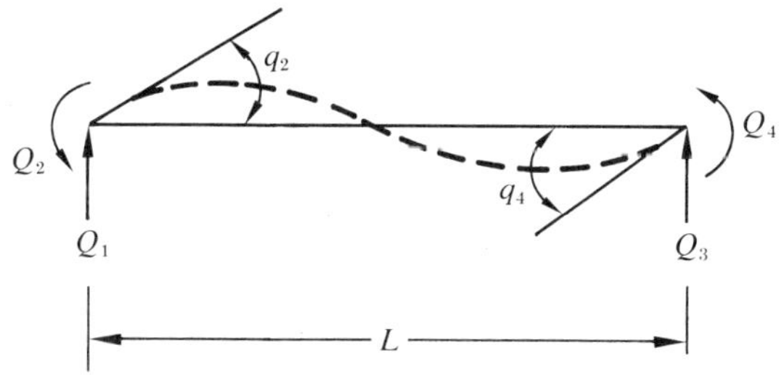

Problem 2.4. Find the total strain energy in the beam structures shown below. Assume constant cross-section.

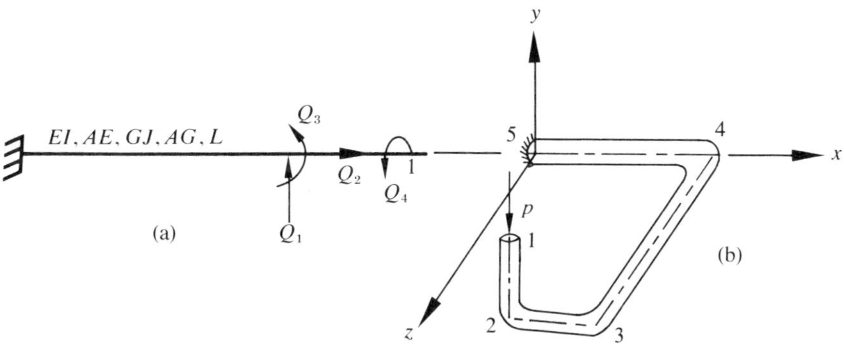

Problem 2.5. Find the strain energy in the rectangular panel shown below.

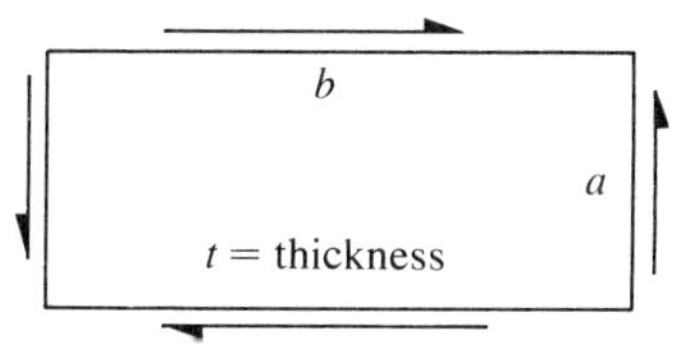

CONSTANT SHEAR FLOW

Problem 2.6. Find the strain energy per unit length of an open cylinder subjected to an external pressure, P_0. The stresses are given by,

$$\sigma_r = \frac{-R_0^2 P_0}{R_0^2 - R_i^2} \frac{(r^2 - R_i^2)}{r^2}$$

$$\sigma_\theta = \frac{-R_0^2 P_0}{R_0^2 - R_i^2} \frac{(r^2 + R_i^2)}{r^2}$$

Shearing stresses = 0. Why?

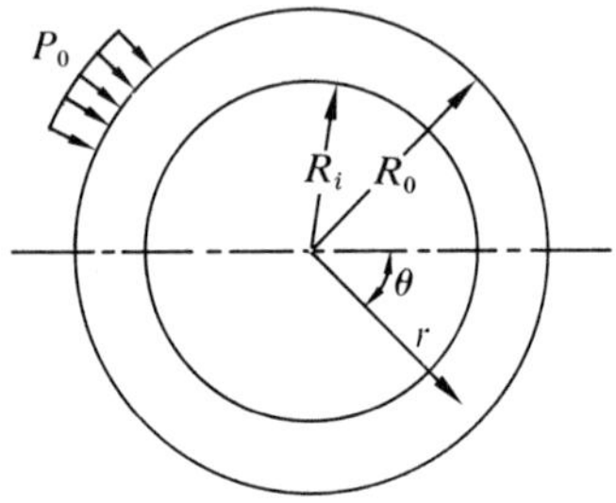

Problem 2.7. Use Castigliano's Theorem to find the slope and deflection of a point, A, of each of the structures shown. Consider only bending stresses.

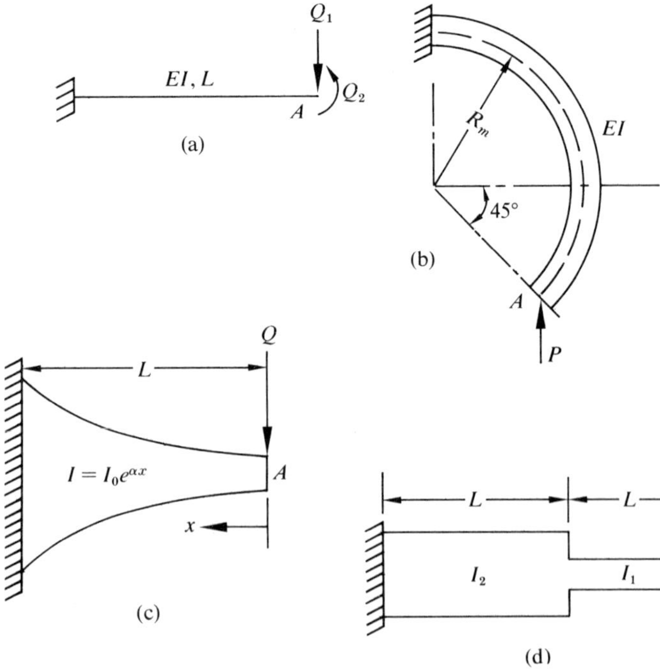

Problem 2.8. Show that Maxwell's Reciprocal Theorem holds for the structure shown.

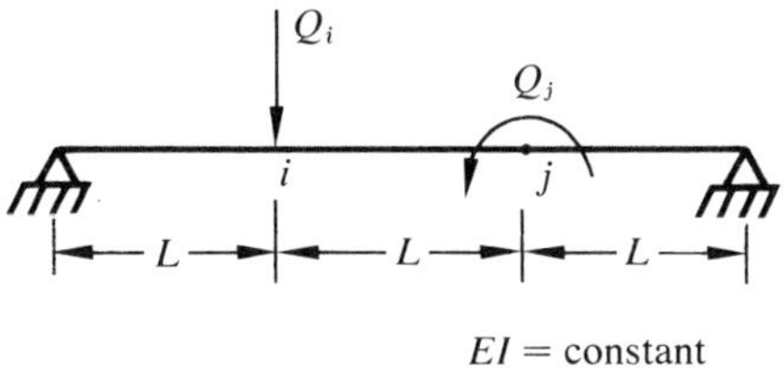

Problem 2.9. The displacement at any point along the length of the axial rod shown below is given by,

$$q_x = \frac{q_j - q_i}{L} x + q_i$$

Find the total strain energy expression for the rod.

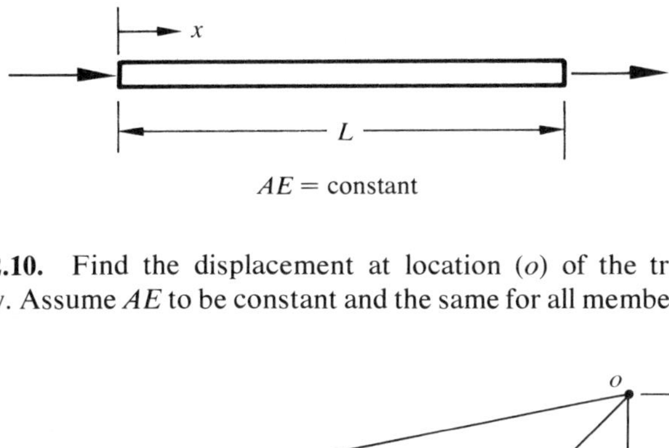

Problem 2.10. Find the displacement at location (o) of the truss structure shown below. Assume AE to be constant and the same for all members.

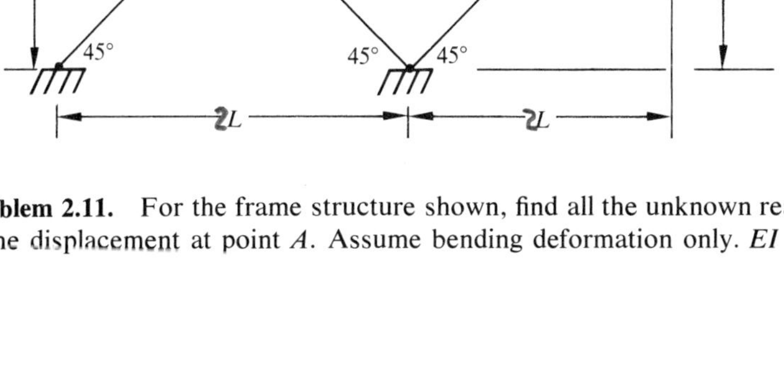

Problem 2.11. For the frame structure shown, find all the unknown reactions and the displacement at point A. Assume bending deformation only. EI = constant.

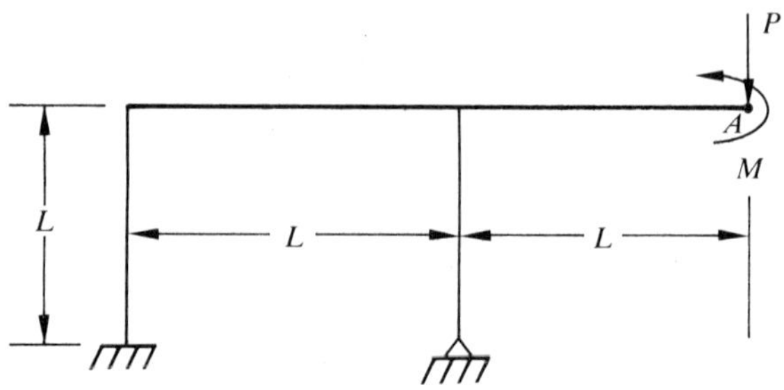

Problem 2.12. Use the method of superposition to find the vertical deflection of point C, of the structures shown.

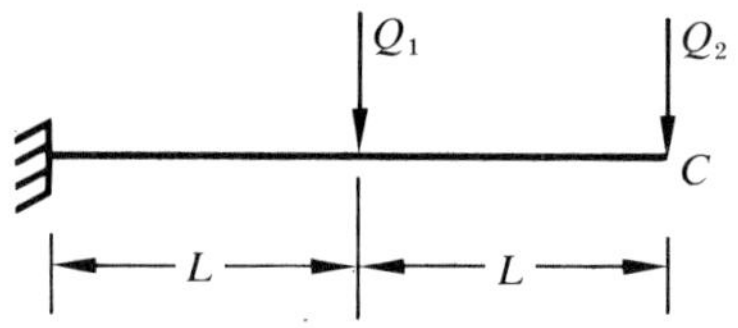

(a) Linear-elastic structure

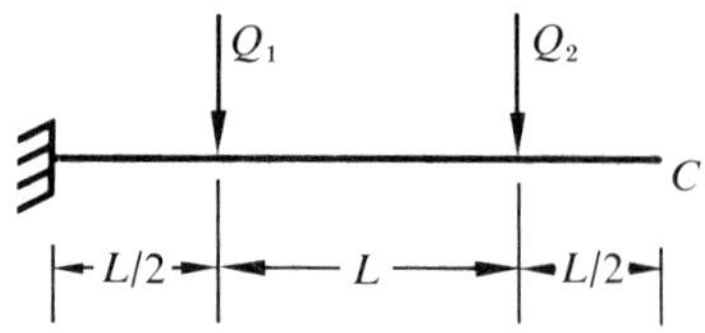

(b) Elastic non-linear structure

Deflection at any point is,

$$y(x) = \frac{Qa^2}{6EI}(2a+3b-3x) \qquad (0 \leqslant x \leqslant b)$$

$$y(x) = \frac{Q}{6EI}(a+b-x)^2(x+2a-b) \qquad (b \leqslant x \leqslant (a+b))$$

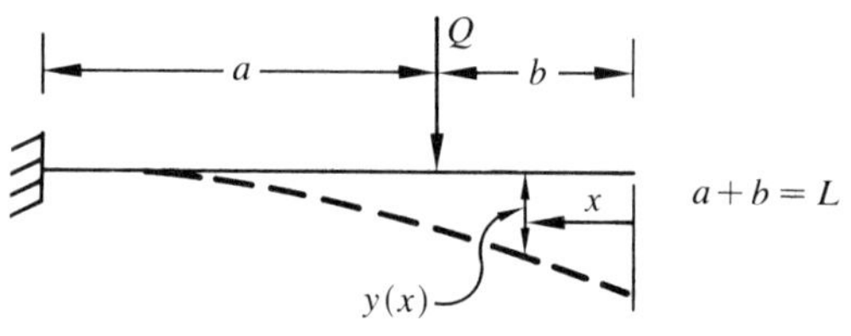

Problem 2.13. Using the strain energy approach, find all reactions on the beam below. Assume EI = constant.

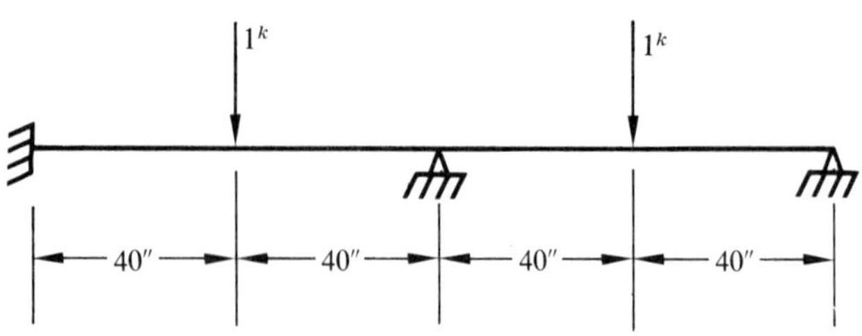

Problem 2.14. Given a quarter circle rigid frame as shown below, find the bending moment expressions at any point along the frame. Use Castigliano's Theorem in your procedure. Neglect shear and axial deformation. Assume EI = constant, and the circle has a radius R. P and M_0 are acting at mid-point between A and B.

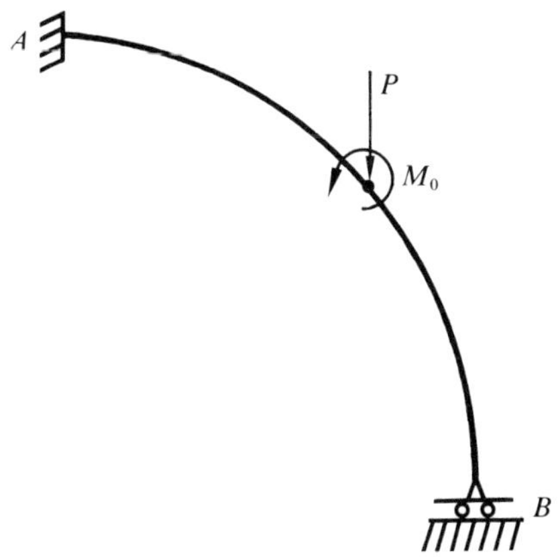

Problem 2.15. Given a column whose behavior is elastic nonlinear subjected to an axial load P. The deflection in the y direction is given by,

$$w = W_0 \sin \frac{\pi x}{L}.$$

The total shortening of the column is $\Delta = (W_0\pi)^2/4L$.

Consider only bending deformation, find the critical load at which the column becomes unstable.

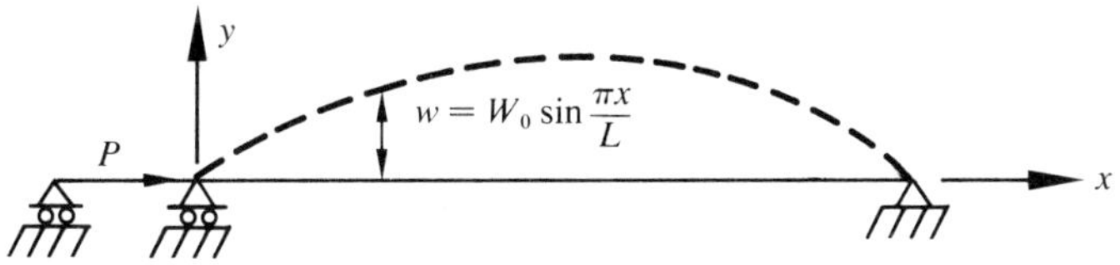

CHAPTER THREE

Direct Stiffness Matrix Method

3.1 INTRODUCTION

The theory and the application of the direct stiffness matrix method to structural analysis are presented in this chapter.

The stiffness matrices of the stiffness coefficients are developed and fully generalized for each of the following structural elements:

Axial rod element
Linear spring element
Torsional rod element
Torsional spring element
Rectilinear beam element
Curved beam element

The method is applied to various structural problems which are common in Civil Engineering, Aerospace and Mechanical Engineering, and other disciplines where such problems exist.

3.2 GENERAL DISCUSSION

Matrix structural analysis is predicated on the ability of the structural analyst to formulate a discrete-element mathematical model equivalent to the real structure. The necessity of structural modeling is to obtain an idealized system with a finite number of degrees of freedom upon which matrix algebra operations can be performed.

In matrix structural analysis the real structure is conceived as an assembly of structural elements (rods, beams, shear panels, etc.) connected at specified node points. The nodes selected are based on the configuration of the real structure as well as on other factors depending on the critical regions to be investigated. (*See* Fig. 3.1 for illustrations.*)

The functional relationship between the nodal forces Q_i (forces acting at the nodes) and their corresponding displacements (nodal displacements) forms the basis of the stiffness matrix approach. In its most

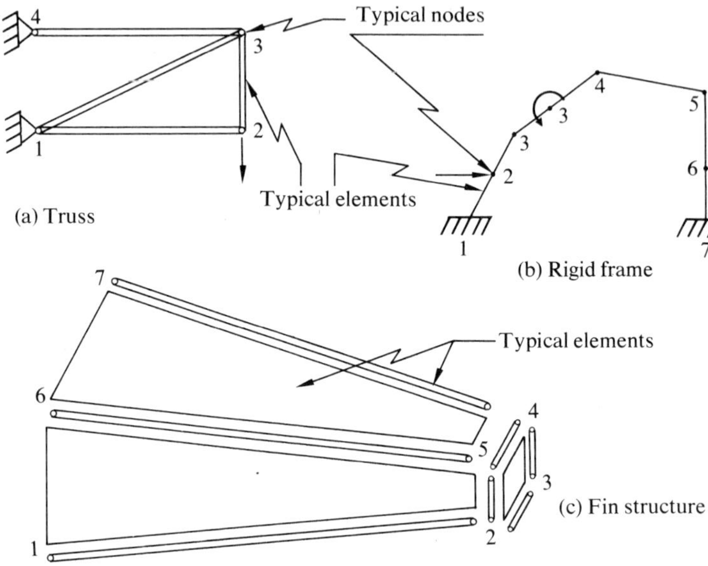

Fig. 3.1. Structural elements.

generalized form, this relationship can be written as,

$$\begin{bmatrix} Q_1 \\ Q_2 \\ Q_3 \\ \cdots \\ Q_i \\ \cdots \\ Q_j \end{bmatrix} = \begin{bmatrix} S_{11} & S_{12} & S_{13} \cdots S_{1i} \cdots S_{1j} \\ S_{21} & S_{22} & S_{23} \cdots S_{2i} \cdots S_{2j} \\ S_{31} & S_{32} & S_{33} \cdots S_{3i} \cdots S_{3j} \\ \cdots & \cdots & \cdots \quad \cdots \quad \cdots \\ S_{i1} & S_{i2} & S_{i3} \cdots S_{ii} \cdots S_{ij} \\ \cdots & \cdots & \cdots \quad \cdots \quad \cdots \\ S_{j1} & S_{j2} & S_{j3} \cdots S_{ji} \cdots S_{jj} \end{bmatrix} \begin{bmatrix} q_1 \\ q_2 \\ q_3 \\ \cdots \\ q_i \\ \cdots \\ q_j \end{bmatrix} \tag{3.1}$$

In a shorthand matrix notation Eq. (3.1) can be expressed by,

$$\{Q\} = [S]\{q\}, \tag{3.1a}$$

in which $\{Q\}$ and $\{q\}$ are the generalized nodal force and displacement

*Reproduced by courtesy of Structural Analysis Associates, Tulsa, Okla.

PRAYER TOWER

ACTUAL

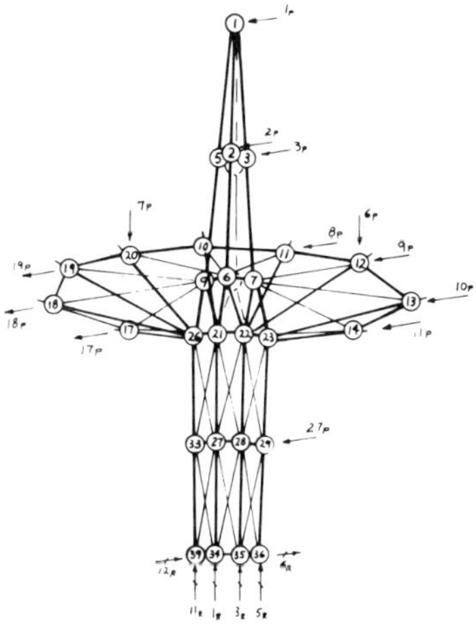

IDEALIZED

HIGH VOLTAGE TOWER

ACTUAL

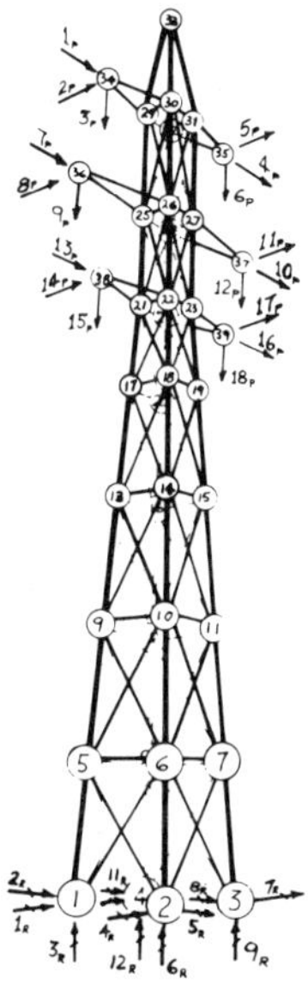

IDEALIZED

(*Courtesy of* Structural Analysis Assocts., Tulsa, Okla.)

BRIDGE

ACTUAL

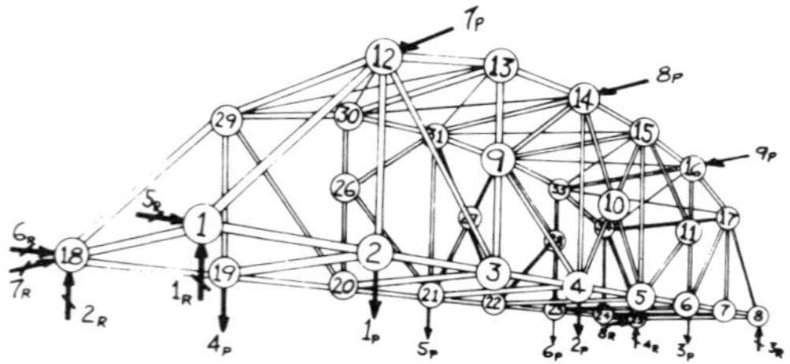

IDEALIZED

INDUSTRIAL COMPLEX

ACTUAL

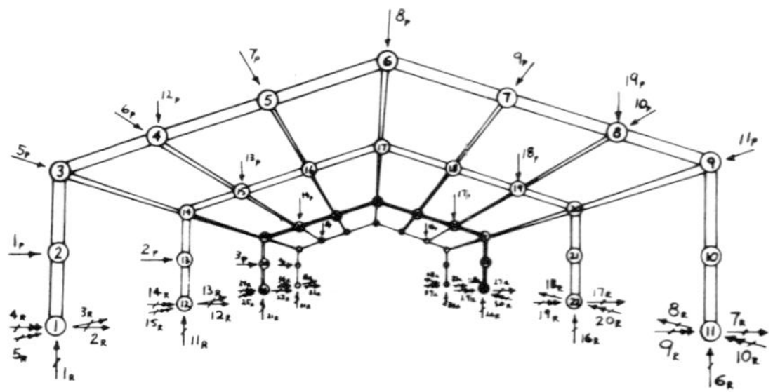

IDEALIZED

(*Courtesy of* Structural Analysis Assocts., Tulsa, Okla.)

CHURCH

ACTUAL

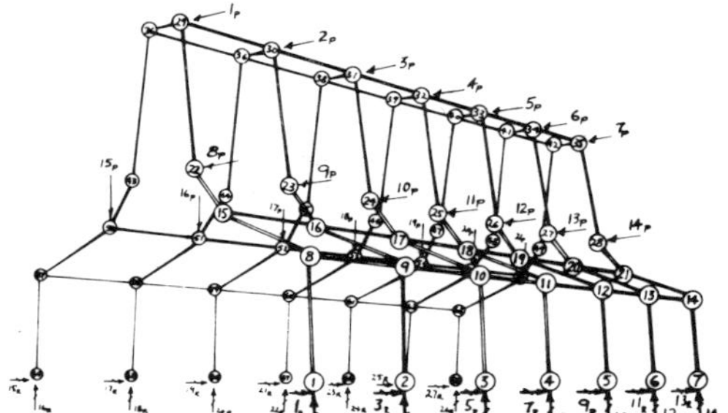

IDEALIZED

WAREHOUSE

ACTUAL

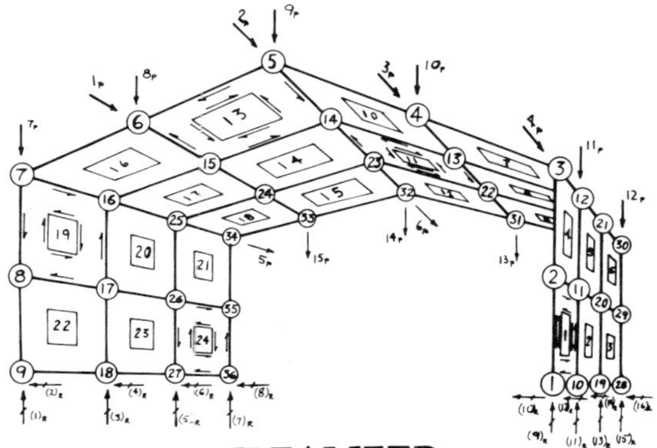

IDEALIZED

(*Courtesy of* Structural Analysis Assocts., Tulsa, Okla.)

vectors respectively; and $[S]$ is the stiffness matrix of the stiffness coefficients. The order of $[S]$ is dictated by the number of degrees of freedom being considered multiplied by the number of nodes present. The stiffness coefficients $S_{ij}(i=j)$ are usually referred to as the main-diagonal elements of the matrix, $[S]$ and are always positive. On the other hand, the coefficients $S_{ij}(i \neq j)$ are called the off-diagonal elements of the matrix. From Maxwell's Reciprocal Theorem $S_{ij} = S_{ji}$, hence the matrix $[S]$ is symmetric about the main diagonal.

By definition, a stiffness coefficient S_{ij} is a force produced at i due to a unit displacement at j. Hence, the problem of deriving the elements of the stiffness matrix is handled systematically by giving the jth coordinate a unit displacement, holding all other coordinates at zero displacements. The resulting forces at all coordinates due to this unit displacement form the stiffness coefficients of the jth column of the stiffness matrix $[S]$. This can be readily seen by letting the coordinate displacement q_j be the only non-zero displacement in Eq. (3.1). Thus, expanding the equation yields,

$$
\begin{aligned}
Q_i &= S_{1j}q_j \\
Q_2 &= S_{2j}q_j \\
&\cdots \\
&\cdots \\
&\cdots \\
Q_i &= S_{ij}q_j \\
&\cdots \\
&\cdots \\
&\cdots \\
Q_j &= S_{jj}q_j
\end{aligned}
$$

If we divide the above equations each by q_j and then let q_j be unity, then it is apparent the stiffness coefficients of the jth column are obtained.

Therefore, we can state that the matrix $[S]$ can be fully populated (all its coefficients found) by giving each coordinate (one at a time) a unit displacement holding all others zero and evaluating the resulting forces at all coordinates. (*See* Fig. 3.2 for illustrations.)

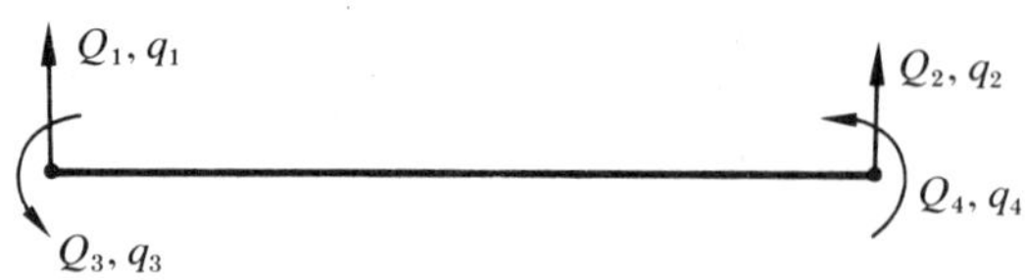

(a) Beam with four coordinate forces and displacements: 2 translations, 2 rotations.

Fig. 3.2. Beam element nodal coordinate displacements.

(b) Unit translation: $q_1 = 1$

(c) Unit translation: $q_2 = 1$

(d) Unit ~~translation~~ rotation: $q_3 = 1$

(e) Unit ~~translation~~ rotation: $q_4 = 1$

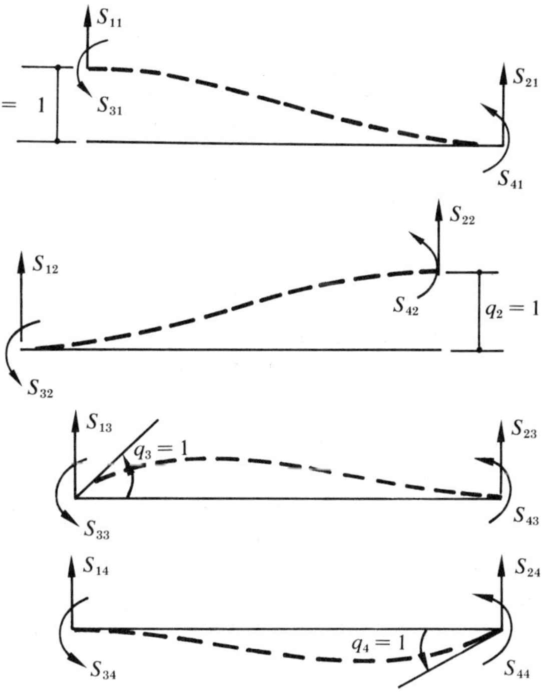

Fig. 3.2. Beam element nodal coordinate displacements.

3.3 AGGREGATE AND ELEMENT STIFFNESS MATRICES

In view of the complexity of most structures, the amount of work involved in deriving directly the overall stiffness matrix of the structure is often prohibitive. Therefore, it is necessary to have at our disposal individual stiffness matrices (element stiffness matrices) and then by superposing these element matrices obtain the overall stiffness matrix (aggregate stiffness matrix) for the structure. Let us demonstrate this by considering the beam structure shown in Fig. 3.3. For simplicity, assume that the beam is idealized by two structural elements connected at node 2. From Eq. (3.1) the individual matrix equations for elements (a) and (b) can be readily written as,

Element (a) (a) Element (b) (b)

$$\begin{bmatrix} Q_1 \\ Q_2 \\ Q_3 \\ Q_4 \end{bmatrix} = \begin{bmatrix} A_{11} & & \text{Sym.} & \\ A_{21} & A_{22} & & \\ A_{31} & A_{32} & A_{33} & \\ A_{41} & A_{42} & A_{43} & A_{44} \end{bmatrix} \begin{bmatrix} q_1 \\ q_2 \\ q_3 \\ q_4 \end{bmatrix}, \quad \begin{bmatrix} Q_3 \\ Q_4 \\ Q_5 \\ Q_6 \end{bmatrix} = \begin{bmatrix} B_{33} & & \text{Sym.} & \\ B_{43} & B_{44} & & \\ B_{53} & B_{54} & B_{55} & \\ B_{63} & B_{64} & B_{65} & B_{66} \end{bmatrix} \begin{bmatrix} q_3 \\ q_4 \\ q_5 \\ q_6 \end{bmatrix}$$

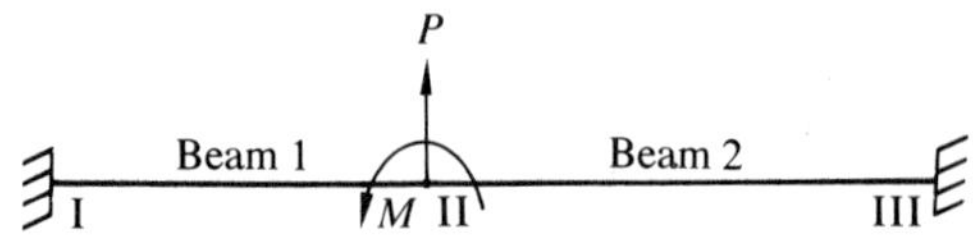

(a) Idealized Structure.

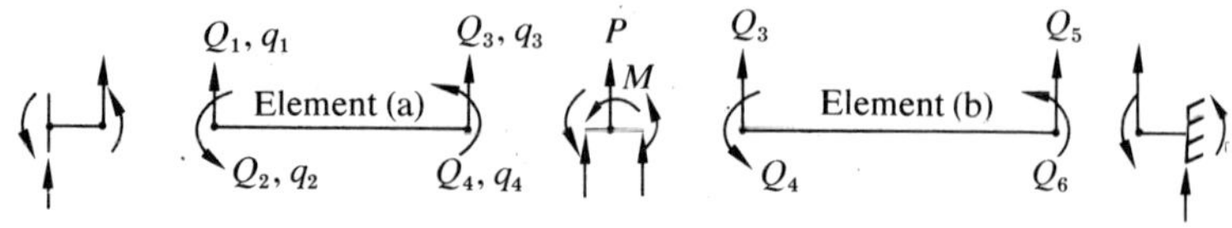

(b) Disassembled Structure.

Fig. 3.3. Beam structure assembly.

in which Q_i $(i = 1\text{–}6)$ is the internal force acting on the element and q_i $(i = 1\text{–}6)$ is the nodal displacement (actual displacement) of the structure. Notice that the order of the stiffness matrices in Eqs. (a) and (b) are determined on the basis that each element has two degrees of freedom per node with maximum of two nodes for each element.

Superposing Eqs. (a) and (b), which is equivalent to assembling the structure back together, yields the aggregate stiffness matrix equation.

$$\begin{bmatrix} Q_1 \\ Q_2 \\ \hline P \\ M \\ \hline Q_5 \\ Q_6 \end{bmatrix} = \left[\begin{array}{cc|cc|cc} A_{11} & & & \text{Sym.} & & \\ A_{21} & A_{22} & & & & \\ \hline A_{31} & A_{32} & A_{33}+B_{33} & & & \\ A_{41} & A_{42} & A_{43}+B_{43} & A_{44}+B_{44} & & \\ \hline 0 & 0 & B_{53} & B_{54} & B_{55} & \\ 0 & 0 & B_{63} & B_{64} & B_{65} & B_{66} \end{array}\right] \begin{bmatrix} 0 \\ 0 \\ \hline q_3 \\ q_4 \\ \hline 0 \\ 0 \end{bmatrix} \tag{c}$$

Again, notice the order of the matrix in Eq. (c) is 6×6, which follows from the fact that the original structure has three nodes with two degrees of freedom (one rotation and one translation) per node. Q_1, Q_2, Q_5, and Q_6 are now the unknown reactions; P and M are the external known applied forces; q_3 and q_4 are the unknown displacements of node 2; and q_1 and q_2, q_5 and q_6 are zero from the boundary conditions of the structure. Notice that the overlap of the two individual element matrices occurs at the common node 2 of the elements.

By matrix partitioning, the following can be extracted from Eq. (c),

$$\begin{bmatrix} P \\ M \end{bmatrix} = \begin{bmatrix} A_{33}+B_{33} & \text{Sym.} \\ A_{43}+B_{43} & A_{44}+B_{44} \end{bmatrix} \begin{bmatrix} q_3 \\ q_4 \end{bmatrix} \tag{d}$$

The stiffness matrix which now appears in Eq. (d) will be referred to as the aggregate reduced stiffness matrix.

Equation (d) can now be solved for the unknown displacements, q_3, and q_4, by matrix inversion. Once these nodal displacements are determined, all the internal forces can be evaluated by utilizing the individual element matrix Eqs. (a) and (b).

In summary, we can state the following:

1. The force vector in the aggregate stiffness matrix equation always represents external applied forces and support reactions.
2. The force vector in the individual element matrix equation always represents the internal forces acting on that element.
3. The displacement vector in both the aggregate and the element stiffness matrix equations always represents the true displacements of the nodes on the structure.

3.4 SIGN CONVENTION

Figure 3.4 shows the sign convention which is adopted throughout the text. Forces and displacements are positive if their vectors point in the direction of the positive coordinates shown.

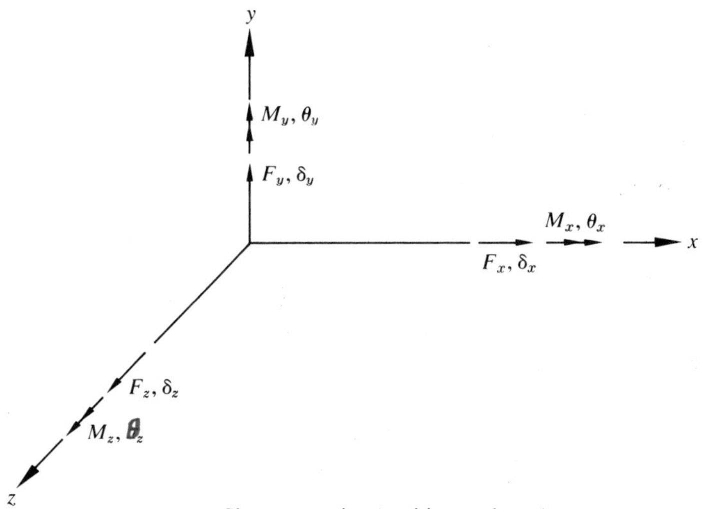

Fig. 3.4. Sign convention (positive as shown).

3.5 AXIAL ROD ELEMENT STIFFNESS MATRIX

Shown in Fig. 3.5 is an axial rod element of uniform cross-section. Based on the definition of an axial rod, the coordinate forces $F_i^{\bar{x}}$ and $F_j^{\bar{x}}$ only, and their corresponding coordinate displacements $\delta_i^{\bar{x}}$ and $\delta_j^{\bar{x}}$ are assumed to exist at nodes i and j, respectively. In our subsequent discussion node i will be referred to as the low node point, and node j as the high node point. For example, in Fig. (b) below node 2 is the low node while nodes 3 and 4 are the high nodes of the elements.

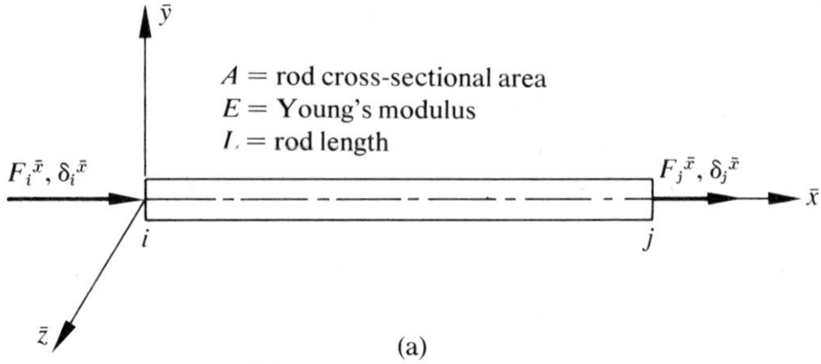

(a)

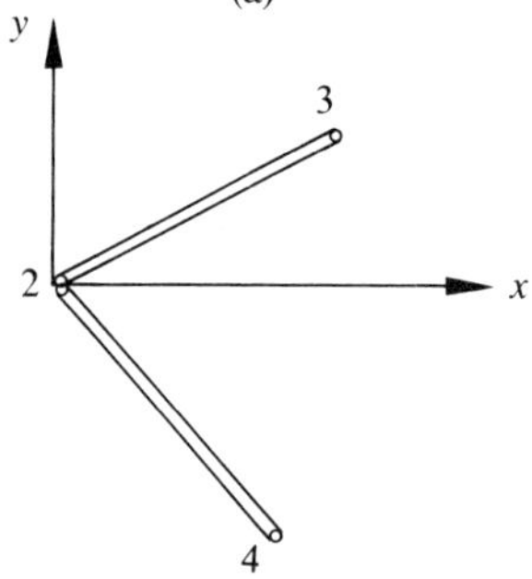

(b) Rod elements showing the high and low node points.

Fig. 3.5. Axial rod element.

The functional relationship between the forces and corresponding displacements of the rod shown, can be easily deduced from Eq. (3.1) as,

$$\begin{bmatrix} Q_1 \\ Q_2 \end{bmatrix} = \begin{bmatrix} S_{11} & S_{12} \\ S_{21} & S_{22} \end{bmatrix} \begin{bmatrix} q_1 \\ q_2 \end{bmatrix} \tag{3.2}$$

where for convenience, the following substitutions are made,

$$F_i^{\bar{x}}, \delta_i^{\bar{x}} = Q_1, q_1 \quad \text{and} \quad F_j^{\bar{x}}, \delta_j^{\bar{x}} = Q_2, q_2.$$

Also the bars indicate that forces and displacements are related in a local coordinate system $\bar{x}\bar{y}$.

The stiffness matrix of Eq. (3.2) is of the order 2×2. Since there are two distinct displacements q_1 and q_2 (Fig. 3.5), it indicates that the order of the element stiffness matrix is governed by the number of distinct nodal coordinate displacements, or as was mentioned before, the order of the matrix is given by the number of degrees of freedom multipled by the total number of nodes.

The stiffness coefficients of the stiffness matrix (Eq. 3.2) will now be derived using the procedure of Section 3.2 (Fig. 3.2).

Condition 1. Let the nodal coordinate displacement q_1, be given a unit displacement while holding all others at zero. This condition is equivalent to fixing the rod as shown in Fig. 3.6. The strain energy in the rod (*see* Eq. 2.32) is given by,

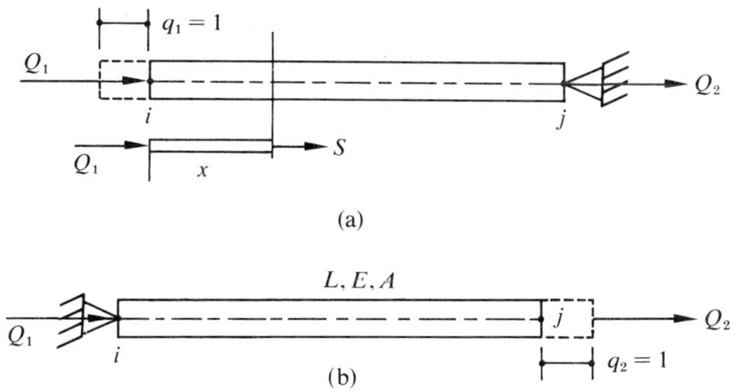

Fig. 3.6. Unit coordinate displacements at nodes *i* and *j*.

$$U = \tfrac{1}{2}\int_0^L \frac{S^2 dx}{AE} \tag{a}$$

where AE = extensional rigidity of the rod,
A = rod area,
E = Young's Modulus.

The axial force S at any distance x is given by,

$$S = -Q_1. \tag{b}$$

Upon substituting Eq. (b) into (a) and integrating yields,

$$U = \frac{Q_1{}^2 L}{2AE}. \tag{c}$$

From Castigliano's Second Theorem (Eq. 2.10) the displacement q_1 is,

$$q_1 = 1 = \frac{\partial U}{\partial Q_1} = \frac{Q_1 L}{AE},$$

therefore,

$$\frac{Q_1}{q_1{=}1} = S_{11} = \frac{AE}{L}, \tag{d}$$

from equilibrium,

$$Q_2 = -Q_1,$$

therefore,

$$\frac{Q_2}{q_1^{=1}} = S_{21} = -\frac{AE}{L}. \tag{e}$$

Condition 2. Let the nodal coordinate displacement q_2 be given a unit displacement while holding all others at zero (Fig. 3.6b).

Following the same procedure as in condition 1, it can be easily shown that,

$$\frac{Q_2}{q_2^{=1}} = S_{22} = \frac{AE}{L}, \tag{f}$$

and

$$\frac{Q_1}{q_2^{=1}} = S_{12} = -\frac{AE}{L}. \tag{g}$$

Hence Eq. (3.2) becomes,

$$\begin{bmatrix} F_i^{\bar{x}} \\ \\ F_j^{\bar{x}} \end{bmatrix} = \begin{bmatrix} \dfrac{AE}{L} & -\dfrac{AE}{L} \\ -\dfrac{AE}{L} & \dfrac{AE}{L} \end{bmatrix} \begin{bmatrix} \delta_i^{\bar{x}} \\ \\ \delta_j^{\bar{x}} \end{bmatrix} \tag{3.2a}$$

Or in shorthand matrix notation Eq. (3.2a) becomes,

$$\{\bar{Q}\} = [\bar{S}]\{\bar{q}\}. \tag{3.2b}$$

The stiffness matrix in Eq. (3.2a) is singular as can be seen easily. This is due to the fact that only rigid body displacement is taking place in the rod and thus there is an infinite set of such displacements which produce the forces $F_i^{\bar{x}}$ and $F_j^{\bar{x}}$.

To render the stiffness matrix nonsingular and ensure a solution, sufficient restraints, i.e. zero displacements prescribed at the boundaries, must be provided. For instance, if end (j) of the rod shown in Fig. 3.6 is held against translation, then Eq. (3.2a) becomes,

$$\begin{bmatrix} F_i^{\bar{x}} \\ F_j^{\bar{x}} \end{bmatrix} = \frac{AE}{L}\begin{bmatrix} 1 & -1 \\ -1 & 1 \end{bmatrix} \begin{bmatrix} \delta_i^{\bar{x}} \\ 0 \end{bmatrix} \tag{a}$$

Deleting the rows and columns corresponding to zero displacement in the above equation yields,

$$F_i^{\bar{x}} = \frac{AE}{L}\delta_i^{\bar{x}},$$

or

$$\delta_i^{\bar{x}} = \frac{F_i^{\bar{x}}L}{AE},$$

which is the displacement of a rod subjected to an axial force $F_i^{\bar{x}}$.

It is appropriate at this time to mention that structural element stiffness matrices may be obtained in a completely different manner than that presented above. The basis of the approach is that the strain energy must be expressed in terms of the nodal coordinate displacements of the structural element under consideration, i.e.,

$$U = U(q_1, q_2, \ldots, q_n).$$

Once the strain energy is formulated in terms of the nodal coordinate displacements, the stiffness coefficients are obtained from the following well known relationship,

$$S_{ij} = \frac{\partial^2 U}{\partial q_i \partial q_j} = \frac{\partial^2 U}{\partial q_j \partial q_i}. \tag{3.3}$$

To illustrate the use of Eq. (3.3), let us consider the rod of Fig. 3.5 again.

The strain energy for an axial rod in terms of strains alone (*see* Eq. 2.31) is given by,

$$U = \tfrac{1}{2} \iiint (E)(\epsilon_{xx})^2 dxdydz, \tag{3.4}$$

in which

$$\iint dydz = \int_A dA = A = \text{area of rod},$$

and

$$\epsilon_{xx} = \text{strain} = \frac{\partial q_x}{\partial x}.$$

Hence Eq. (3.4) becomes,

$$U = \frac{AE}{2} \int_0^L \left(\frac{\partial q}{\partial x}\right)^2 dx. \tag{3.4a}$$

For a uniform rod, the strain is a constant, therefore, we can assume that the displacement q is a linear function of x, i.e.,

$$q = A_1 x + A_2, \tag{a}$$

in which A_1 and A_2 are constants to be determined from boundary conditions.

For the rod under consideration, the following boundary conditions exist,

$$\text{at } x = 0 \qquad q = q_1 \tag{b}$$

and

$$x = L \qquad q = q_2 \tag{c}$$

Utilizing conditions (b) and (c) in Eq. (a) yields,

$$A_2 = q_1$$

and

$$A_1 = \frac{q_2 - q_1}{L}.$$

Therefore, the displacement at any point x along the rod is given by,

$$q = \frac{(q_2 - q_1)x + q_1}{L}. \tag{3.5}$$

Substituting Eq. (3.5) into (3.4a) and integrating yields,

$$U = \frac{AE}{2L}(q_2{}^2 - 2q_1q_2 + q_1{}^2), \tag{3.6}$$

which is the strain energy in the rod expressed in terms of the nodal coordinate displacements.

Using Eqs. (3.3) and (3.6), the following stiffness coefficients are obtained,

$$S_{11} = \frac{\partial^2 U}{\partial q_i{}^2} = \frac{AE}{L}$$

$$S_{12} = S_{21} = \frac{\partial^2 U}{\partial q_1 \partial q_2} = -\frac{AE}{L}$$

and

$$S_{22} = \frac{\partial^2 U}{\partial q_2{}^2} = \frac{AE}{L}$$

which are the same results as those obtained using Castigliano's Second Theorem.

3.6 COORDINATE TRANSFORMATION MATRIX

The stiffness matrix in Eq. (3.2a) is limited in its application to rod elements that are oriented as shown in Fig. 3.5. To accommodate for elements which are arbitrarily oriented in space, a coordinate transformation matrix is developed. This matrix allows for two rigid body rotations of the rod. (*See* Fig. 3.7.)

a. Rotation about the y-axis in the xz plane;
b. Rotation about the z_1-axis in the x_1y_1 plane.

The relationship between $x_1y_1z_1$ and xyz coordinate axes can easily be

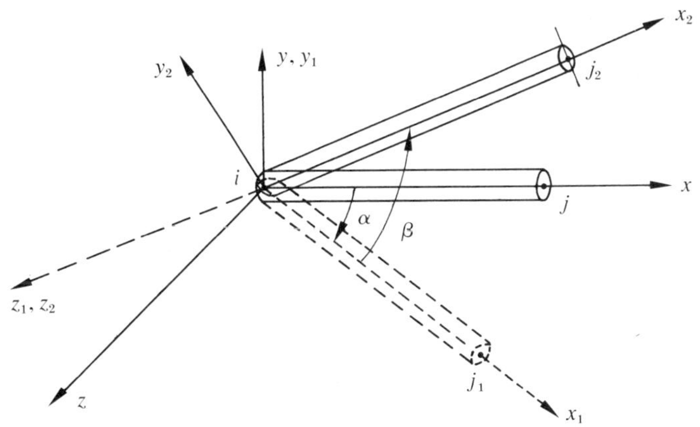

Fig. 3.7. Coordinate transformation.

derived and is given by,

$$\begin{bmatrix} x_1 \\ y_1 \\ z_1 \end{bmatrix} = \begin{bmatrix} \cos\alpha & 0 & \sin\alpha \\ 0 & 1 & 0 \\ -\sin\alpha & 0 & \cos\alpha \end{bmatrix} \begin{bmatrix} x \\ y \\ z \end{bmatrix} \tag{3.7}$$

Likewise, the relationship between $x_2y_2z_2$ and $x_1y_1z_1$ is given by,

$$\begin{bmatrix} x_2 \\ y_2 \\ z_2 \end{bmatrix} = \begin{bmatrix} \cos\beta & \sin\beta & 0 \\ -\sin\beta & \cos\beta & 0 \\ 0 & 0 & 1 \end{bmatrix} \begin{bmatrix} x_1 \\ y_1 \\ z_1 \end{bmatrix} \tag{3.8}$$

in which α and β are always measured from a horizontal line drawn through the low node point as shown in Fig. 3.7.

Upon substituting Eq. (3.7) into (3.8) and performing matrix multiplication, the following matrix equation is obtained,

$$\begin{bmatrix} x_2 \\ y_2 \\ z_2 \end{bmatrix} = \begin{bmatrix} \cos\alpha\cos\beta & \sin\beta & \cos\beta\sin\alpha \\ -\cos\alpha\sin\beta & \cos\beta & -\sin\alpha\sin\beta \\ -\sin\alpha & 0 & \cos\alpha \end{bmatrix} \begin{bmatrix} x \\ y \\ z \end{bmatrix} \tag{3.9}$$

or, in shorthand matrix notation,

$$\{\bar{\Gamma}\} = [T]\{\Gamma\}, \tag{3.9a}$$

where $[T]$ is referred to as the transformation matrix; $\bar{\Gamma}$ = local coordinate axes; and Γ = reference coordinate axes.

Hence, in lieu of Eq. (3.9a), forces and displacements in the local coordinate axes can be related to those in the reference coordinate axes as follows,

$$\{\bar{Q}\} = [T]\{Q\}, \tag{3.10}$$

and

$$\{\bar{q}\} = [T]\{q\}, \tag{3.11}$$

where $[T]$ is given in Eq. (3.9).

If a stiffness matrix is generated based on the local axes of the element (centroidal principal axes), like for instance in Section 3.5, then the matrix equation is given by,

$$\{\bar{Q}\} = [\bar{S}]\{\bar{q}\}. \tag{3.12}$$

If Eqs. (3.10) and (3.11) are substituted into Eq. (3.12) the following is obtained,

$$[T]\{Q\} = [\bar{S}][T]\{q\},$$

or premultiplying by the inverse of $[T]$ yields:

$$\{Q\} = [T]^{-1}[\bar{S}][T]\{q\}, \tag{3.13}$$

or

$$\{Q\} = [S]\{q\}. \tag{3.13a}$$

Where $[S]$ now is the generalized stiffness matrix for an arbitrarily oriented element.

$$[S] = [T]^{-1}[\bar{S}][T]. \tag{3.14}$$

The transformation matrix in Eq. (3.9) can also be expressed in terms of the element node point coordinates. From Fig. 3.8 the following rela-

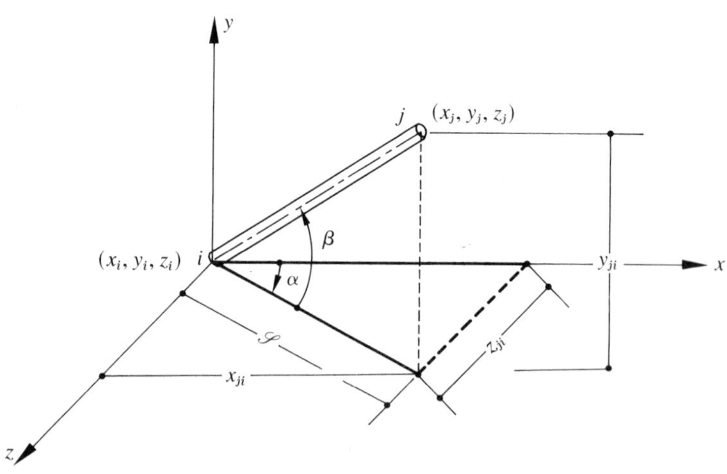

Fig. 3.8. Coordinate transformation.

tions can be easily obtained,

$$\sin\alpha = \frac{z_{ji}}{\mathscr{S}}, \qquad \cos\alpha = \frac{x_{ji}}{\mathscr{S}}$$

$$\sin\beta = \frac{y_{ji}}{L}, \qquad \cos\beta = \frac{\mathscr{S}}{L}$$

in which

$$z_{ji} = z_j - z_i, \qquad x_{ji} = x_j - x_i, \qquad y_{ji} = y_j - y_i, \qquad \mathscr{S} = (z_{ji}^2 + x_{ji}^2)^{1/2},$$

and L = length of element $= (x_{ji}^2 + y_{ji}^2 + z_{ji}^2)^{1/2}$.

Therefore, the transformation matrix in terms of the coordinates of the nodes can be written as,

$$[T] = \begin{bmatrix} \frac{x_{ji}}{L} & \frac{y_{ji}}{L} & \frac{z_{ji}}{L} \\ -\frac{x_{ji}y_{ji}}{L\mathscr{S}} & \frac{\mathscr{S}}{L} & -\frac{y_{ji}z_{ji}}{L\mathscr{S}} \\ -\frac{z_{ji}}{\mathscr{S}} & 0 & \frac{x_{ji}}{\mathscr{S}} \end{bmatrix} \tag{3.15a}$$

It is obvious that if x_{ji} and z_{ji} are both zero, then $\mathscr{S}$ is undefined, which makes Eq. (3.15a) meaningless. Hence, for this special case Eq. (3.15a) is given by,

$$[T] = \begin{bmatrix} 0 & 1 & 0 \\ -1 & 0 & 0 \\ 0 & 0 & 1 \end{bmatrix} \qquad \text{(For } y_{ji} \text{ positive)} \tag{3.15b}$$

and

$$[T] = \begin{bmatrix} 0 & -1 & 0 \\ 1 & 0 & 0 \\ 0 & 0 & 1 \end{bmatrix} \qquad \text{(For } y_{ji} \text{ negative)} \tag{3.15c}$$

Therefore, on the basis of Eq. (3.14) the generalized stiffness matrix can be obtained for the axial rod element oriented as shown in Fig. 3.9. Notice that the rod in the local coordinate system (xyz) has all zero forces and displacements except those acting along its neutral axis.

To cope with the number of coordinate forces and displacements in the reference coordinate system, Eq. (3.2a) can be rewritten as follows,

$$\begin{bmatrix} F_i^{\bar{x}} \\ F_i^{\bar{y}} \\ F_i^{\bar{z}} \\ F_j^{\bar{x}} \\ F_j^{\bar{y}} \\ F_j^{\bar{z}} \end{bmatrix} = \frac{AE}{L} \begin{bmatrix} 1 & 0 & 0 & -1 & 0 & 0 \\ 0 & 0 & 0 & 0 & 0 & 0 \\ 0 & 0 & 0 & 0 & 0 & 0 \\ -1 & 0 & 0 & 1 & 0 & 0 \\ 0 & 0 & 0 & 0 & 0 & 0 \\ 0 & 0 & 0 & 0 & 0 & 0 \end{bmatrix} \begin{bmatrix} \delta_i^{\bar{x}} \\ \delta_i^{\bar{y}} \\ \delta_i^{\bar{z}} \\ \delta_j^{\bar{x}} \\ \delta_j^{\bar{y}} \\ \delta_j^{\bar{z}} \end{bmatrix} \tag{3.16}$$

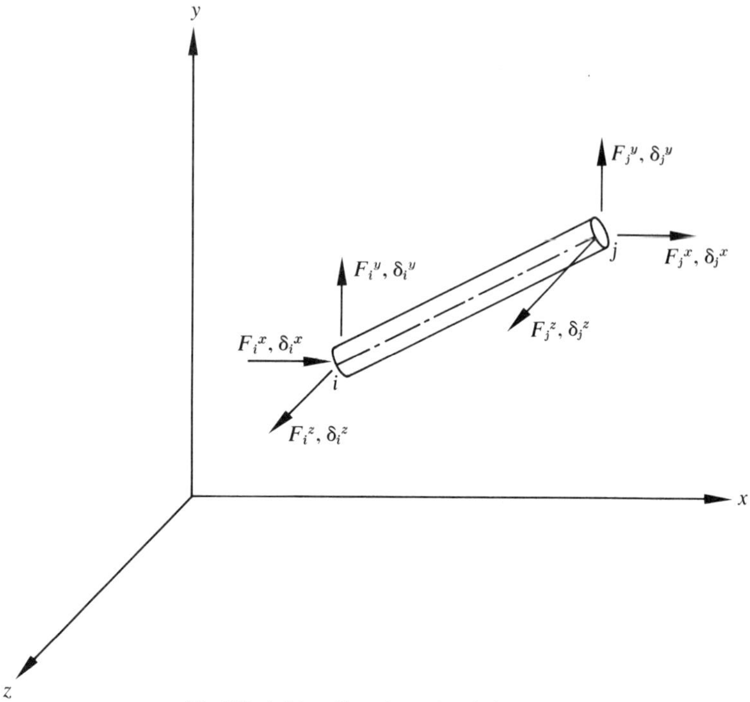

Fig. 3.9. Arbitrarily oriented rod element.

Substituting $[\bar{S}]$ from Eq. (3.16), $[T]$ and its inverse from Eq. (3.9) into Eq. (3.14) we have,

$$\frac{AE}{L}\begin{bmatrix} T_{11} & T_{21} & T_{31} & 0 & 0 & 0 \\ T_{12} & T_{22} & T_{32} & 0 & 0 & 0 \\ T_{13} & T_{23} & T_{33} & 0 & 0 & 0 \\ 0 & 0 & 0 & T_{11} & T_{21} & T_{31} \\ 0 & 0 & 0 & T_{12} & T_{22} & T_{32} \\ 0 & 0 & 0 & T_{13} & T_{23} & T_{33} \end{bmatrix}\begin{bmatrix} 1 & 0 & 0 & -1 & 0 & 0 \\ 0 & 0 & 0 & 0 & 0 & 0 \\ 0 & 0 & 0 & 0 & 0 & 0 \\ -1 & 0 & 0 & 1 & 0 & 0 \\ 0 & 0 & 0 & 0 & 0 & 0 \\ 0 & 0 & 0 & 0 & 0 & 0 \end{bmatrix}$$

$$\times\begin{bmatrix} T_{11} & T_{12} & T_{13} & 0 & 0 & 0 \\ T_{21} & T_{22} & T_{23} & 0 & 0 & 0 \\ T_{31} & T_{32} & T_{33} & 0 & 0 & 0 \\ 0 & 0 & 0 & T_{11} & T_{12} & T_{13} \\ 0 & 0 & 0 & T_{21} & T_{22} & T_{23} \\ 0 & 0 & 0 & T_{31} & T_{32} & T_{33} \end{bmatrix} = [S] \qquad \text{(a)}$$

Performing the multiplication in Eq. (a) yields the following generalized stiffness matrix for an axial rod element,

$$[S] = \frac{AE}{L} \times \begin{bmatrix} T_{11}^2 & & & \text{Sym.} & & \\ T_{11}T_{12} & T_{12}^2 & & & & \\ T_{11}T_{13} & T_{12}T_{13} & T_{13}^2 & & & \\ -T_{11}^2 & -T_{11}T_{12} & -T_{11}T_{13} & T_{11}^2 & & \\ -T_{11}T_{12} & -T_{12}^2 & -T_{12}T_{13} & T_{11}T_{12} & T_{12}^2 & \\ -T_{11}T_{13} & -T_{12}T_{13} & -T_{13}^2 & T_{11}T_{13} & T_{12}T_{13} & T_{13}^2 \end{bmatrix} \tag{3.16a}$$

and Eq. (3.13) in an expanded form becomes,

$$\begin{bmatrix} F_i^x \\ F_i^y \\ F_i^z \\ F_j^x \\ F_j^y \\ F_j^z \end{bmatrix} = \frac{AE}{L} \begin{bmatrix} T_{11}^2 & & & \text{Sym.} & & \\ T_{11}T_{12} & T_{12}^2 & & & & \\ T_{11}T_{13} & T_{12}T_{13} & T_{13}^2 & & & \\ -T_{11}^2 & -T_{11}T_{12} & -T_{11}T_{13} & T_{11}^2 & & \\ -T_{11}T_{12} & -T_{12}^2 & -T_{12}T_{13} & T_{11}T_{12} & T_{12}^2 & \\ -T_{11}T_{13} & -T_{12}T_{13} & -T_{13}^2 & T_{11}T_{13} & T_{12}T_{13} & T_{13}^2 \end{bmatrix} \begin{bmatrix} \delta_i^x \\ \delta_i^y \\ \delta_i^z \\ \delta_j^x \\ \delta_j^y \\ \delta_j^z \end{bmatrix} \tag{3.17}$$

The stiffness matrix of Eq. (3.17) is singular. To enhance a solution of the Eq. (3.17) sufficient boundary conditions must be imposed.

Illustrative Examples

Example 1

Shown in the figure below is a pin-ended truss-structure. Using the stiffness matrix method, find the nodal displacements and the internal forces in each member.

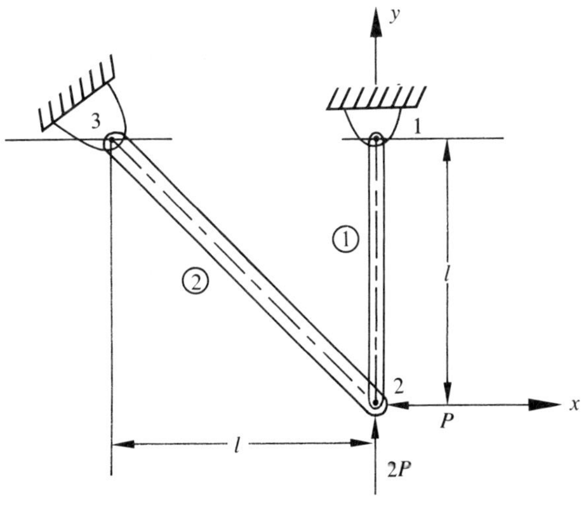

Given: Area of member ② $= \sqrt{2} =$ (area of member ①) and E is the same for both members.

Solution. First of all we formulate the element stiffness matrix for each rod. Since the elements are arbitrarily oriented with respect to the xy axis we need their respective transformation coefficients. Also, notice that this is a planar type truss, hence the z-components do not appear in the equations. The coordinates of the nodes are,

Node	x	y
1	0	l
2	0	0
3	$-l$	l

From Eq. (3.15) the coefficients of the transformation matrices for rod 1 and 2 are;

Rod 1 (Eq. 3.15c),

$$T_{11} = 0,$$

$$T_{12} = \frac{y_{21}}{(x_{21}^2 + y_{21}^2 + z_{21}^2)^{1/2}} = -1. \qquad \text{(a)}$$

Rod 2 (Eq. 3.15a)

$$T_{11} = \frac{x_{32}}{(x_{32}^2 + y_{32}^2 + z_{32}^2)^{1/2}} = -\frac{1}{\sqrt{2}}, \qquad \text{(b)}$$

$$T_{12} = \frac{y_{32}}{(x_{32}^2 + y_{32}^2 + z_{32}^2)^{1/2}} = \frac{1}{\sqrt{2}}.$$

Deleting rows and columns in Eq. (3.17) corresponding to the z-components and then substituting (a) and (b), yields,

Rod 1

$$\begin{bmatrix} F_1^x \\ F_1^y \\ F_2^x \\ F_2^y \end{bmatrix} = \frac{AE}{l} \begin{bmatrix} 0 & & \text{Sym.} & \\ 0 & 1 & & \\ 0 & 0 & 0 & \\ 0 & -1 & 0 & 1 \end{bmatrix} \begin{bmatrix} \delta_1^x \\ \delta_1^y \\ \delta_2^x \\ \delta_2^y \end{bmatrix} \qquad \text{(c)}$$

Rod 2

$$\begin{bmatrix} F_2^x \\ F_2^y \\ F_3^x \\ F_3^y \end{bmatrix} = \frac{AE}{l\sqrt{2}} \begin{bmatrix} 0.5 & & \text{Sym.} & \\ -0.5 & 0.5 & & \\ -0.5 & 0.5 & 0.5 & \\ 0.5 & -0.5 & -0.5 & 0.5 \end{bmatrix} \begin{bmatrix} \delta_2^x \\ \delta_2^y \\ \delta_3^x \\ \delta_3^y \end{bmatrix} \qquad \text{(d)}$$

The boundary conditions are,

Node 1

$$\delta_1{}^x = \delta_1{}^y = 0,$$

Node 3

$$\delta_3{}^x = \delta_3{}^y = 0.$$

The external applied forces are,

Node 2

$$F_2{}^x = -P,$$

$$F_2{}^y = 2P.$$

The aggregate stiffness matrix is obtained by superposing the individual matrices of rod 1 and 2, or

$$\frac{AE}{l}\begin{bmatrix} 0 & & & & & \\ 0 & 1 & & \text{Sym.} & & \\ 0 & 0 & 0 & & & \\ 0 & -1 & 0 & 1 & & \\ 0 & 0 & 0 & 0 & 0 & \\ 0 & 0 & 0 & 0 & 0 & 0 \end{bmatrix} + \frac{AE}{l}\begin{bmatrix} 0 & & & & & \\ 0 & 0 & & & \text{Sym.} & \\ 0 & 0 & 0.353 & & & \\ 0 & 0 & -0.353 & 0.353 & & \\ 0 & 0 & -0.353 & 0.353 & 0.353 & \\ 0 & 0 & 0.353 & -0.353 & -0.353 & 353 \end{bmatrix}$$

$$= \frac{AE}{l}\begin{bmatrix} 0 & & & & & \\ 0 & 1 & & & \text{Sym.} & \\ 0 & 0 & 0.353 & & & \\ 0 & -1 & -0.353 & 1.353 & & \\ 0 & 0 & -0.353 & 0.353 & 0.353 & \\ 0 & 0 & 0.353 & -0.353 & -0.353 & 0.353 \end{bmatrix} \qquad \text{(e)}$$

Notice that we have added rows and corresponding columns of zeros in the individual matrices before they were superposed, the reason being that since there are three nodes to the given structure with two degrees of freedom at each node, the size of the aggregate stiffness matrix would be on the order of 6 × 6. Hence, to be able to add the individual matrices and make it compatible with the expected size of the resulting aggregate matrix, the zeros of rows and columns corresponding to the nodes which do not exist on the rod under consideration, have to be added. For instance, in our example, rod 1 has the nodes 1 and 2, and therefore we add rows and columns of zeros corresponding to node 3. Likewise rod 2 has the nodes 2 and 3, and thus we should add rows and columns corresponding to node 1.

The aggregate stiffness matrix equation for the structure now becomes

$$\begin{bmatrix} F_1^x \\ F_1^y \\ -P \\ 2P \\ F_3^x \\ F_3^y \end{bmatrix} = \frac{AE}{l} \begin{bmatrix} 0 & & & & & \\ 0 & 1 & & \text{Sym.} & & \\ 0 & 0 & 0.353 & & & \\ 0 & -1 & -0.353 & 1.353 & & \\ 0 & 0 & -0.353 & 0.353 & 0.353 & \\ 0 & 0 & 0.353 & -0.353 & -0.353 & 0.353 \end{bmatrix} \begin{bmatrix} 0 \\ 0 \\ \delta_2^x \\ \delta_2^y \\ 0 \\ 0 \end{bmatrix} \tag{f}$$

In Eq. (f), δ_2^x and δ_2^y are the only unknown displacements, while F_1^x, F_1^y, F_3^x and F_3^y are the unknown reactions. By matrix partitioning or by deleting rows and corresponding columns which correspond to zero displacements, the following can be extracted

$$\begin{bmatrix} -P \\ 2P \end{bmatrix} = \frac{AE}{l} \begin{bmatrix} 0.353 & -0.353 \\ -0.353 & 1.353 \end{bmatrix} \begin{bmatrix} \delta_2^x \\ \delta_2^y \end{bmatrix} \tag{g}$$

Inverting the stiffness matrix in Eq. (g) and solving for the displacements yield

$$\delta_2^x = -\frac{1.83\,Pl}{AE}, \qquad \delta_2^y = \frac{Pl}{AE}. \tag{h}$$

Substituting (h) into (c) and (d) the following internal loads are found,
For rod 1

$$F_1^x = 0, \qquad F_1^y = -P, \qquad F_2^x = 0, \qquad F_2^y = P,$$

For rod 2

$$F_2^x = -P, \qquad F_1^y = P, \qquad F_3^x = P, \qquad F_3^y = -P.$$

Example 2

Find the internal forces in all members and the node deflections of the truss-structure shown below (p. 69). Assume that the area of each rod is 1 in^2 and the modulus of elasticity is 29×10^6 psi.

Node Coordinates			Rod Length	
Node	x	y	Rod	Length
1	0	0	1	$20\sqrt{2}$
2	40	0	2	$20\sqrt{2}$
3	20	20	3	$20\sqrt{10}$
4	80	40	4	$40\sqrt{2}$

Coefficients of Transformation Matrix

Rod	Coefficients	
1	$T_{11} = T_{12} = 1/\sqrt{2}$	
2	$T_{11} = -1/\sqrt{2}$,	$T_{12} = 1/\sqrt{2}$
3	$T_{11} = 3/\sqrt{10}$,	$T_{12} = 1/\sqrt{10}$
4	$T_{11} = 1/\sqrt{2}$,	$T_{12} = T_{11}$

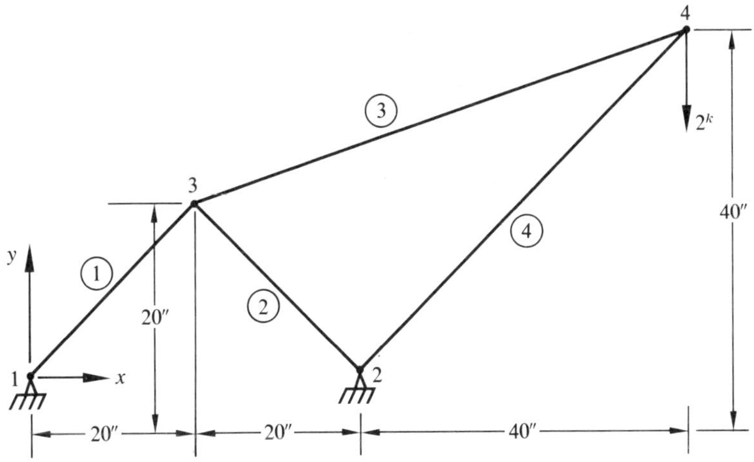

Individual Element Stiffness Matrices

Rod 1

$$\frac{AE}{20\sqrt{2}}\begin{bmatrix} 0.5 & & \text{Sym.} & \\ 0.5 & 0.5 & & \\ -0.5 & -0.5 & 0.5 & \\ -0.5 & -0.5 & 0.5 & 0.5 \end{bmatrix};$$

Rod 2

$$\frac{AE}{20\sqrt{2}}\begin{bmatrix} 0.5 & & \text{Sym.} & \\ -0.5 & 0.5 & & \\ -0.5 & 0.5 & 0.5 & \\ 0.5 & -0.5 & -0.5 & 0.5 \end{bmatrix}$$

Rod 3

$$\frac{AE}{20\sqrt{10}}\begin{bmatrix} 0.9 & & \text{Sym.} & \\ 0.3 & 0.1 & & \\ -0.9 & -0.3 & 0.9 & \\ -0.3 & -0.1 & 0.3 & 0.1 \end{bmatrix};$$

Rod 4

$$\frac{AE}{40\sqrt{2}}\begin{bmatrix} 0.5 & & \text{Sym.} & \\ 0.5 & 0.5 & & \\ -0.5 & -0.5 & 0.5 & \\ -0.5 & -0.5 & 0.5 & 0.5 \end{bmatrix}$$

Aggregate Stiffness Matrix

$$\begin{bmatrix} F_1^x \\ F_1^y \\ F_2^x \\ F_2^y \\ F_3^x \\ F_3^y \\ F_4^x \\ F_4^y \end{bmatrix} = \frac{AE}{20\sqrt{2}}\begin{bmatrix} 0.5 & & & & & & & \\ 0.5 & 0.5 & & & \text{Sym.} & & & \\ 0.0 & 0.0 & 0.75 & & & & & \\ 0.0 & 0.0 & -0.25 & 0.75 & & & & \\ -0.5 & -0.5 & -0.5 & 0.5 & 1.4 & & & \\ -0.5 & -0.5 & 0.5 & -0.5 & 0.13 & 1.04 & & \\ 0.0 & 0.0 & -0.25 & -0.25 & -0.4 & -0.13 & 0.65 & \\ 0.0 & 0.0 & -0.25 & -0.25 & -0.13 & -0.04 & 0.38 & 0.29 \end{bmatrix}\begin{bmatrix} \delta_1^x \\ \delta_1^y \\ \delta_2^x \\ \delta_2^y \\ \delta_3^x \\ \delta_3^y \\ \delta_4^x \\ \delta_4^y \end{bmatrix}$$

Boundary Conditions

$$\delta_1{}^x = \delta_1{}^y = \delta_2{}^x = \delta_2{}^y = 0$$

Applied Loads: $F_4{}^y = -2000\# = -2$ Kips

Hence, the reduced stiffness matrix equation is

$$\begin{bmatrix} 0 \\ 0 \\ 0 \\ -2^k \end{bmatrix} = \frac{AE}{20\sqrt{2}} \begin{bmatrix} 1.4 & 0.13 & -0.4 & -0.14 \\ 0.13 & 1.04 & -0.13 & -0.04 \\ -0.4 & -0.13 & 0.65 & 0.38 \\ -0.13 & -0.04 & 0.38 & 0.29 \end{bmatrix} \begin{bmatrix} \delta_3{}^x \\ \delta_3{}^y \\ \delta_4{}^x \\ \delta_4{}^y \end{bmatrix}$$

The unknown displacements, now, can be evaluated by inverting the reduced stiffness matrix. Once these displacements are known the internal forces on each member can be found from the individual element stiffness matrix equations.

Example 3

Find the internal forces of all members and node deflections of the truss structure shown. Assume that $AE = 30 \times 10^6$ for all members.

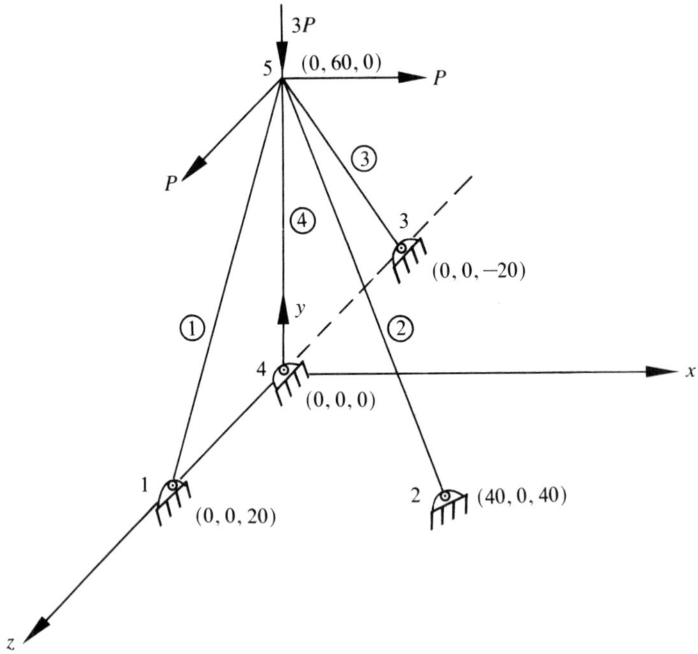

Node Coordinates

Node	x	y	z
1	0	0	20
2	40	0	20
3	0	0	−20
4	0	0	0
5	0	60	0

Rod Length

Rod	Length
1	63.2
2	74.8
3	63.2
4	60.0

Coefficients of Transformation Matrix

Rod	Coefficients		
1	$T_{11}=0,$	$T_{12}=0.95,$	$T_{13}=-0.32$
2	$T_{11}=-0.53,$	$T_{12}=0.8,$	$T_{13}=-0.26$
3	$T_{11}=0,$	$T_{12}=0.95,$	$T_{13}=0.32$
4	$T_{11}=0,$	$T_{12}=1,$	$T_{13}=0$

Individual Element Stiffness Matrices

Rod 1

$$\frac{AE}{20\sqrt{2}}\begin{bmatrix} 0 & & & & & \\ 0 & 0.4 & & \text{Sym.} & & \\ 0 & -0.14 & 0.046 & & & \\ 0 & 0 & 0 & 0 & & \\ 0 & -0.4 & 0.14 & 0 & 0.4 & \\ 0 & 0.14 & -0.046 & 0 & -0.14 & 0.046 \end{bmatrix}$$

Rod 2

$$\frac{AE}{20\sqrt{2}}\begin{bmatrix} 0.11 & & & & & \\ -0.16 & 0.14 & & & \text{Sym.} & \\ 0.052 & -0.079 & 0.026 & & & \\ -0.11 & 0.16 & -0.052 & 0.11 & & \\ 0.16 & -0.24 & 0.079 & -0.16 & 0.24 & \\ -0.052 & 0.079 & -0.026 & 0.052 & -0.079 & 0.026 \end{bmatrix}$$

Rod 3

$$\frac{AE}{20\sqrt{2}}\begin{bmatrix} 0 & & & & & \\ 0 & 0.4 & & \text{Sym.} & & \\ 0 & 0.14 & 0.046 & & & \\ 0 & 0 & 0 & 0 & & \\ 0 & -0.4 & -0.14 & 0 & 0.4 & \\ 0 & -0.14 & -0.046 & 0 & 0.14 & 0.046 \end{bmatrix}$$

Rod 4

$$\frac{AE}{20\sqrt{2}}\begin{bmatrix} 0 & & & & & \\ 0 & 0.22 & & \text{Sym.} & & \\ 0 & 0 & 0 & & & \\ 0 & 0 & 0 & 0 & & \\ 0 & -0.22 & 0 & 0 & 0.22 & \\ 0 & 0 & 0 & 0 & 0 & 0 \end{bmatrix}$$

The Aggregate Stiffness Matrix is,

$$
\begin{bmatrix} F_1^x \\ F_1^y \\ F_1^z \\ F_2^x \\ F_2^y \\ F_2^z \\ F_3^x \\ F_3^y \\ F_3^z \\ F_4^x \\ F_4^y \\ F_4^z \\ F_5^x \\ F_5^y \\ F_5^z \end{bmatrix}
= \frac{AE}{20\sqrt{2}}
\begin{bmatrix}
0 & 0 & 0 & & & & & & & & & & 0 & 0 & 0 \\
0 & 0.4 & -0.14 & & & & & & & & & & 0 & -0.4 & 0.14 \\
0 & -0.14 & 0.046 & & & & & & & & & & 0 & 0.14 & -0.046 \\
 & & & 0.11 & -0.16 & 0.052 & & & & & & & -0.11 & 0.16 & -0.052 \\
 & & & -0.16 & 0.24 & -0.079 & & & & & & & 0.16 & -0.24 & 0.079 \\
 & & & 0.052 & 0.079 & 0.026 & & & & & & & -0.052 & 0.079 & -0.026 \\
 & & & & & & 0 & 0 & 0 & & & & 0 & 0 & 0 \\
 & & & & & & 0 & 0.4 & 0.14 & & & & 0 & -0.4 & -0.14 \\
 & & & & & & 0 & 0.14 & 0.046 & & & & 0 & -0.14 & -0.046 \\
 & & & & & & & & & 0 & 0 & 0 & 0 & 0 & 0 \\
 & & & & & & & & & 0 & 0.22 & 0 & 0 & -0.22 & 0 \\
 & & & & & & & & & 0 & 0 & 0 & 0 & 0 & 0 \\
0 & 0 & 0 & -0.11 & 0.16 & -0.052 & 0 & 0 & 0 & 0 & 0 & 0 & 0.11 & -0.16 & 0.052 \\
0 & -0.4 & 0.14 & 0.16 & -0.24 & 0.079 & 0 & -0.4 & -0.14 & 0 & -0.22 & 0 & -0.16 & 1.26 & -0.079 \\
0 & 0.14 & -0.046 & -0.052 & 0.079 & -0.026 & 0 & -0.14 & -0.046 & 0 & 0 & 0 & 0.052 & -0.079 & 0.118
\end{bmatrix}
\begin{bmatrix} \delta_1^x \\ \delta_1^y \\ \delta_1^z \\ \delta_2^x \\ \delta_2^y \\ \delta_2^z \\ \delta_3^x \\ \delta_3^y \\ \delta_3^z \\ \delta_4^x \\ \delta_4^y \\ \delta_4^z \\ \delta_5^x \\ \delta_5^y \\ \delta_5^z \end{bmatrix}
$$

The boundary conditions are

$$\delta_1^x = \delta_1^y = \delta_1^z = 0$$

$$\delta_2^x = \delta_2^y = \delta_2^z = 0$$

$$\delta_3^x = \delta_3^y = \delta_3^z = 0$$

$$\delta_4^x = \delta_4^y = \delta_4^z = 0$$

The applied loads are

$$F_5^x = P, \qquad F_5^y = -3P, \qquad F_5^z = P.$$

Therefore, the reduced aggregate stiffness matrix equation is,

$$\begin{bmatrix} P \\ -3P \\ P \end{bmatrix} = \frac{AE}{20\sqrt{2}} \begin{bmatrix} 0.11 & -0.16 & 0.052 \\ -0.16 & 1.26 & -0.079 \\ 0.052 & -0.079 & 0.12 \end{bmatrix} \begin{bmatrix} \delta_5^x \\ \delta_5^y \\ \delta_5^z \end{bmatrix}$$

The above equation can be easily solved for the displacements and by utilizing the individual element matrix equations the internal forces can be determined.

3.7 LINEAR SPRING STIFFNESS MATRIX

Shown in Fig. 3.10 is a typical linear spring element of stiffness K. Node point coordinate forces $F_i^{\bar{x}}$ and $F_j^{\bar{x}}$ with their corresponding displacements are assumed to exist at nodes i and j. Linear springs can resist only loads which are directed along its own neutral axis. Basically then, a linear spring adheres to the same rules as an axial rod in its force carrying capabilities.

Therefore, the basic constituent matrix equations derived in the

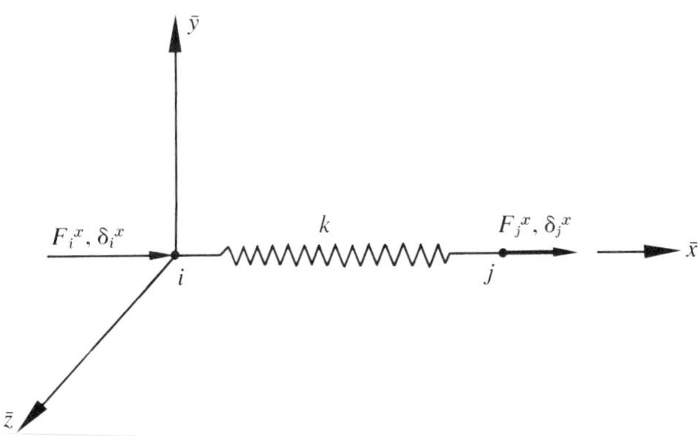

Fig. 3.10. Linear spring element.

previous section apply here if we substitute the stiffness K for AE/L. Hence, no further discussion is necessary in this respect.

Example 4

Given the spring assembly as shown, find all node point displacements and internal forces.

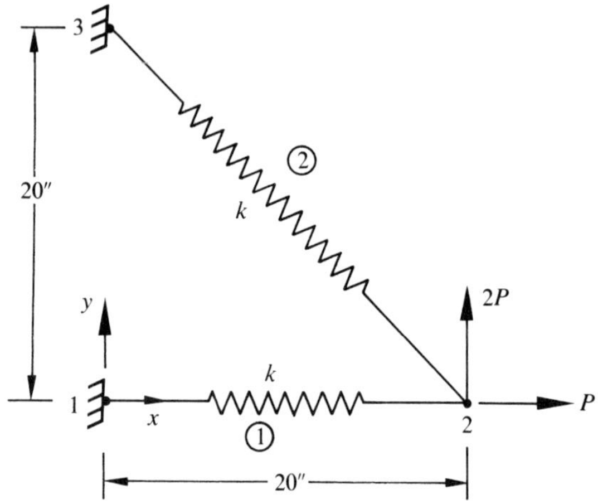

From Eq. (3.17) the stiffness matrix equation for spring 1 and 2 are
Spring 1

$$\begin{bmatrix} F_1{}^x \\ F_1{}^y \\ F_2{}^x \\ F_2{}^y \end{bmatrix} = \begin{bmatrix} K & & \text{Sym.} & \\ 0 & 0 & & \\ -K & 0 & K & \\ 0 & 0 & 0 & 0 \end{bmatrix} \begin{bmatrix} \delta_1{}^x \\ \delta_1{}^y \\ \delta_2{}^x \\ \delta_2{}^y \end{bmatrix} \tag{a}$$

$$\begin{bmatrix} F_2{}^x \\ F_2{}^y \\ F_3{}^x \\ F_3{}^y \end{bmatrix} = \begin{bmatrix} 0.5\,K & & \text{Sym.} & \\ -0.5\,K & 0.5\,K & & \\ -0.5\,K & 0.5\,K & 0.5\,K & \\ 0.5\,K & -0.5\,K & -0.5\,K & 0.5\,K \end{bmatrix} \begin{bmatrix} \delta_2{}^x \\ \delta_2{}^y \\ \delta_3{}^x \\ \delta_3{}^y \end{bmatrix} \tag{b}$$

Superposing Eqs. (a) and (b), introducing the boundary conditions and substituting for the external applied loads yield,

$$\begin{bmatrix} F_1{}^x \\ F_1{}^y \\ P \\ 2P \\ F_3{}^x \\ F_3{}^y \end{bmatrix} = K \left[\begin{array}{cc|cc|cc} 1 & & & \text{Sym.} & & \\ 0 & 0 & & & & \\ -1 & 0 & 1.5 & & & \\ 0 & 0 & -0.5 & 0.5 & & \\ \hline 0 & 0 & -0.5 & 0.5 & 0.5 & \\ 0 & 0 & 0.5 & -0.5 & -0.5 & 0.5 \end{array}\right] \begin{bmatrix} 0 \\ 0 \\ \delta_2{}^x \\ \delta_2{}^y \\ 0 \\ 0 \end{bmatrix} \tag{c}$$

From the partitioned matrix in Eq. (c) the following can be extracted,

$$\begin{bmatrix} P \\ 2P \end{bmatrix} = K\begin{bmatrix} 1.5 & -0.5 \\ -0.5 & 0.5 \end{bmatrix}\begin{bmatrix} \delta_2{}^x \\ \delta_2{}^y \end{bmatrix} \tag{d}$$

Solving for displacements

$$\delta_2{}^x = 3P/K,$$
$$\delta_2{}^y = 7P/K.$$

The internal forces can be calculated from Eqs. (a) and (b) and are shown below

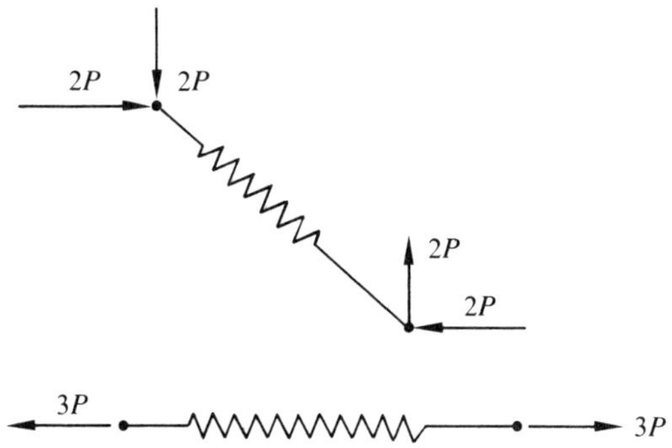

Example 5

Given the structure shown below find the reduced aggregate matrix equation for the structure.

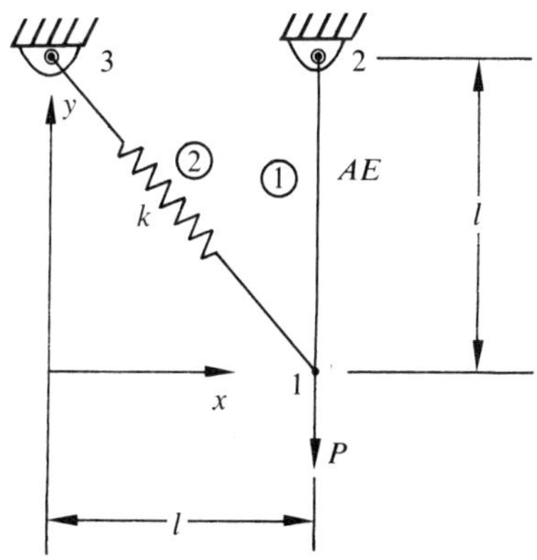

From Eq. (3.17) the stiffness matrices for elements 1 and 2 are

Axial Rod Element 1

$$\begin{bmatrix} F_1{}^x \\ F_1{}^y \\ F_2{}^x \\ F_2{}^y \end{bmatrix} = \begin{bmatrix} 0 & & \text{Sym.} & \\ AE/l & 0 & & \\ 0 & 0 & 0 & \\ -AE/l & 0 & 0 & AE/l \end{bmatrix} \begin{bmatrix} \delta_1{}^x \\ \delta_1{}^y \\ \delta_2{}^x \\ \delta_2{}^y \end{bmatrix} \tag{a}$$

Linear Spring Element 2

$$\begin{bmatrix} F_1{}^x \\ F_1{}^y \\ F_3{}^x \\ F_3{}^y \end{bmatrix} = \begin{bmatrix} 0.5\,K & & \text{Sym.} & \\ -0.5\,K & 0.5\,K & & \\ -0.5\,K & 0.5\,K & 0.5\,K & \\ 0.5\,K & -0.5\,K & -0.5\,K & 0.5\,K \end{bmatrix} \begin{bmatrix} \delta_1{}^x \\ \delta_1{}^y \\ \delta_3{}^x \\ \delta_3{}^y \end{bmatrix} \tag{b}$$

Superposing Eqs. (a) and (b) the following matrix equation for the whole structure is obtained.

$$\begin{bmatrix} 0 \\ -P \\ F_2{}^x \\ F_2{}^y \\ F_3{}^x \\ F_3{}^y \end{bmatrix} = \begin{bmatrix} 0 & & & \text{Sym.} & & \\ AE/l & 0 & & & & \\ 0 & 0 & 0 & & & \\ -AE/l & 0 & 0 & AE/l & & \\ 0 & 0 & 0 & 0 & 0 & \\ 0 & 0 & 0 & 0 & 0 & 0 \end{bmatrix} \begin{bmatrix} \delta_1{}^x \\ \delta_1{}^y \\ \delta_2{}^x \\ \delta_2{}^y \\ \delta_3{}^x \\ \delta_3{}^y \end{bmatrix} +$$

$$\begin{bmatrix} 0.5\,K & & & & & \\ -0.5\,K & 0.5\,K & & \text{Sym.} & & \\ 0 & 0 & 0 & & & \\ 0 & 0 & 0 & 0 & & \\ -0.5\,K & 0.5\,K & 0 & 0 & 0.5\,K & \\ 0.5\,K & -0.5\,K & 0 & 0 & -0.5\,K & 0.5\,K \end{bmatrix} \begin{bmatrix} \delta_1{}^x \\ \delta_1{}^y \\ \delta_2{}^x \\ \delta_2{}^y \\ \delta_3{}^x \\ \delta_3{}^y \end{bmatrix} =$$

$$\begin{bmatrix} 0.5\,K & & & & & \\ -0.5\,K+AE/l & 0.5\,K & & \text{Sym.} & & \\ 0 & 0 & 0 & & & \\ -AE/l & 0 & 0 & AE/l & & \\ -0.5\,K & 0.5\,K & 0 & 0 & 0.5\,K & \\ 0.5\,K & -0.5\,K & 0 & 0 & -0.5\,K & 0.5\,K \end{bmatrix} \begin{bmatrix} \delta_1{}^x \\ \delta_1{}^y \\ \delta_2{}^x \\ \delta_2{}^y \\ \delta_3{}^x \\ \delta_3{}^y \end{bmatrix} \tag{c}$$

Deleting the rows and corresponding columns in Eq. (c) which zero displacements are prescribed yields the following reduced aggregate stiffness matrix equation

$$\begin{bmatrix} 0 \\ -P \end{bmatrix} = \begin{bmatrix} 0.5\,K & \\ -0.5\,K+AE/l & 0.5\,K \end{bmatrix} \begin{bmatrix} \delta_1{}^x \\ \delta_1{}^y \end{bmatrix} \tag{d}$$

3.8 TORSIONAL ROD AND TORSIONAL SPRING STIFFNESS MATRICES

Figure 3.11 shows a torsional rod and spring elements with positive coordinate forces and displacements acting at nodes i and j. The "right-hand-rule" is used here in establishing the direction of the torque. If we assume that only pure torsion can be transmitted through the rod and spring, then the stiffness matrix equations can be readily deduced from Eq. (3.2a).

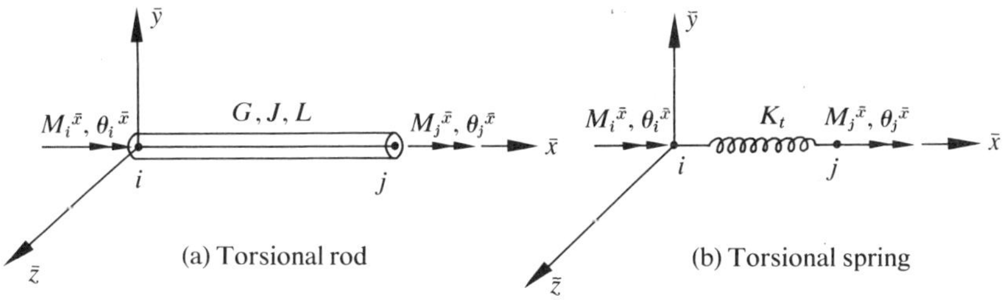

Fig. 3.11. Torsional elements.

Thus, the torsional rod stiffness matrix equation in the local coordinate system is,

$$\begin{bmatrix} M_i^{\bar{x}} \\ M_j^{\bar{x}} \end{bmatrix} = \frac{GJ}{L} \begin{bmatrix} 1 & -1 \\ -1 & 1 \end{bmatrix} \begin{bmatrix} \theta_i^{\bar{x}} \\ \theta_i^{\bar{y}} \end{bmatrix} \tag{3.18}$$

where G = shear modulus,
J = polar moment of inertia,
L = length of rod,
GJ/L = torsional stiffness (lb-in/rad.).

Likewise for the torsional spring

$$\begin{bmatrix} M_i^{\bar{x}} \\ M_j^{\bar{x}} \end{bmatrix} = K_t \begin{bmatrix} 1 & -1 \\ -1 & 1 \end{bmatrix} \begin{bmatrix} \theta_i^{\bar{x}} \\ \theta_j^{\bar{x}} \end{bmatrix} \tag{3.19}$$

where K_t is the torsional spring stiffness.

In a reference coordinate system Eq. (3.18) becomes

$$\begin{bmatrix} M_{ix}^{x} \\ M_{iy}^{x} \\ M_{iz}^{x} \\ M_{jx}^{x} \\ M_{jy}^{x} \\ M_{jz}^{x} \end{bmatrix} = \frac{GJ}{L} \begin{bmatrix} T_{11}^2 & & & & & \\ T_{11}T_{12} & T_{12}^2 & & & \text{Sym.} & \\ T_{11}T_{13} & T_{12}T_{13} & T_{13}^2 & & & \\ -T_{11}^2 & -T_{11}T_{12} & -T_{11}T_{13} & T_{11}^2 & & \\ -T_{11}T_{12} & -T_{12}^2 & -T_{12}T_{13} & T_{11}T_{12} & T_{12}^2 & \\ -T_{11}T_{13} & -T_{12}T_{13} & -T_{13}^2 & T_{11}T_{13} & T_{12}T_{13} & T_{13}^2 \end{bmatrix} \begin{bmatrix} \theta_{ix}^{x} \\ \theta_{iy}^{x} \\ \theta_{iz}^{x} \\ \theta_{jx}^{x} \\ \theta_{jy}^{x} \\ \theta_{jz}^{x} \end{bmatrix} \tag{3.20}$$

where the subscripts x, y, and z indicate components of M^x and θ^x in the x, y, and z directions, respectively, i.e.,

$$M_i^x = [(M_{ix}^x)^2 + (M_{iy}^x)^2 + (M_{iz}^x)^2]^{1/2}$$

and

$$\theta_i^x = [(\theta_{ix}^x)^2 + (\theta_{iy}^x)^2 + (\theta_{iz}^x)^2]^{1/2}$$

Example 6

Find the twist at nodes 1 and 2 of the rod shown below.

Boundary conditions: $\theta_3^x = 0$

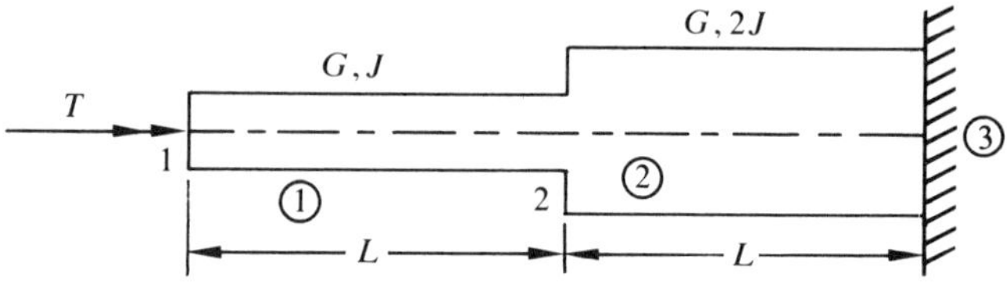

The elements of the transformation matrix are

For 1

$$T_{11} = 1, \qquad T_{12} = 0, \qquad T_{13} = 0.$$

For 2

$$T_{11} = 1, \qquad T_{12} = 0, \qquad T_{13} = 0.$$

Utilizing Eq. (3.20) for rods 1 and 2 and superposing, the following aggregate stiffness matrix equation is obtained

$$\begin{bmatrix} T \\ 0 \\ T_3 \end{bmatrix} = \frac{GJ}{L}\begin{bmatrix} 1 & -1 & 0 \\ -1 & 3 & -2 \\ 0 & -2 & 2 \end{bmatrix}\begin{bmatrix} \theta_1^x \\ \theta_2^x \\ 0 \end{bmatrix} \tag{a}$$

Inverting the stiffness matrix in Eq. (a), after deleting rows and columns corresponding to zero displacements, the angles of twist can be calculated

$$\theta_1^x = 3TL/2GJ \qquad \text{and} \qquad \theta_2^x = TL/2GJ$$

3.9 RECTILINEAR BEAM ELEMENT STIFFNESS MATRIX

Figure 3.12 shows a typical rectilinear beam element. In-plane and normal-to-plane forces and displacements are assumed to exist at nodes i and j. The forces consist of bending, torsional, axial and transverse shearing forces. Forces and corresponding displacements are shown only at node j. Similar expressions exist at node i.

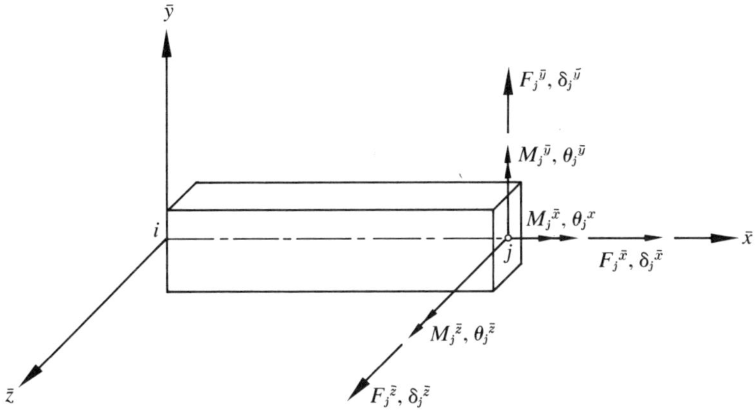

Fig. 3.12. Rectilinear beam element.

If we assume that xy and xz planes coincide with the principal axis of the element cross-section then a change of displacements in one plane (xy plane for instance) induces zero forces in the xz plane. Hence, we can derive the stiffness matrix of the stiffness coefficients in one plane independent of the other.

Thus, the strain energy of the beam element (Fig. 3.12) due to forces in the xy plane can be written as

$$U = \tfrac{1}{2}\int_0^L \left\{\frac{[S(x)]^2}{AE} + \frac{[M(x)]^2}{EI} + \frac{[V(x)]^2}{AG} + \frac{[T(x)]^2}{JG}\right\} dx, \tag{3.21}$$

in which AE, EI, AG, and JG are the extensional, flexural, shear, and torsional rigidities of the beam, respectively.

For convenience, let us use the generalized force and displacement notations given in Eqs. (2.1) and (2.2), respectively.

Thus

$$\begin{bmatrix} Q_1 \\ Q_2 \\ Q_3 \\ Q_4 \\ Q_5 \\ Q_6 \\ Q_7 \\ Q_8 \\ Q_9 \\ Q_{10} \\ Q_{11} \\ Q_{12} \end{bmatrix} ; \begin{bmatrix} q_1 \\ q_2 \\ q_3 \\ q_4 \\ q_5 \\ q_6 \\ q_7 \\ q_8 \\ q_9 \\ q_{10} \\ q_{11} \\ q_{12} \end{bmatrix} = \begin{bmatrix} F_i^x \\ F_i^y \\ F_i^z \\ M_i^x \\ M_i^y \\ M_i^z \\ F_j^x \\ F_j^y \\ F_j^z \\ M_j^x \\ M_j^y \\ M_j^z \end{bmatrix} ; \begin{bmatrix} \delta_i^x \\ \delta_i^y \\ \delta_i^z \\ \theta_i^x \\ \theta_i^y \\ \theta_i^z \\ \delta_j^x \\ \delta_j^y \\ \delta_j^z \\ \theta_j^x \\ \theta_j^y \\ \theta_j^z \end{bmatrix} \tag{2.1; 2.2}$$

Based on the previous assumption, that displacements which are confined to the xz plane (in this case they are $q_1, q_2, q_4, q_6, q_7, q_8, q_{10}$ and q_{12}) induce no forces in the xy plane ($Q_3 = Q_5 = Q_9 = Q_{11} = 0$), we can conclude that the following stiffness coefficients are zero,

$$\left.\begin{aligned} \bar{S}_{3i} &= \frac{Q_3}{q_i} = 0 \\ \bar{S}_{5i} &= \frac{Q_5}{q_i} = 0 \\ \bar{S}_{9i} &= \frac{Q_9}{q_i} = 0 \\ \bar{S}_{11,i} &= \frac{Q_{11}}{q_i} = 0 \end{aligned}\right\} \quad \begin{aligned} &(i = 1, 2, 4, 6, 7, 8, 10, 12) \\ &(q_i = \text{unit displacement}) \end{aligned} \tag{3.22}$$

To derive the rest of the stiffness coefficients we take the partial derivative of Eq. (2.32) with respect to a nonzero coordinate force Q_i ($i = 1, 2, 4, 6, 7, 8, 10, 12$). From Castigliano's Second Theorem this yields a displacement q_i in the direction of Q_i, or

$$q_i = \frac{\partial U}{\partial Q_i} = \int_0^L \left\{ \frac{S}{AE}\frac{\partial S}{\partial Q_i} + \frac{M}{EI}\frac{\partial M}{\partial Q_i} + \frac{V}{AG}\frac{\partial V}{\partial Q_i} + \frac{T}{JG}\frac{\partial T}{\partial Q_i} \right\} dx, \tag{3.23}$$

where S, M, V, and T are the internal forces at any section along the beam elemented and can be expressed in terms of Q_i as below.

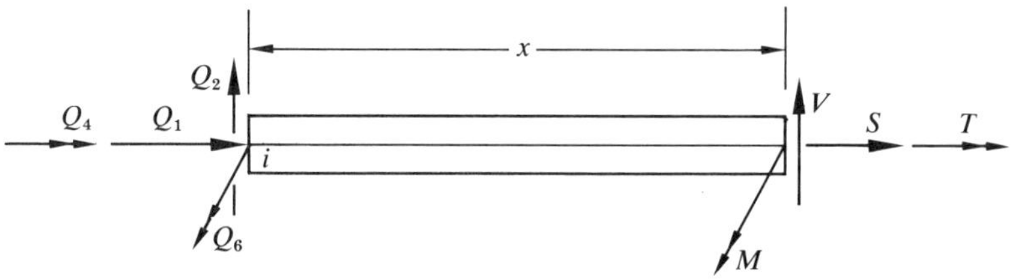

First consider a section x distance from node i in Fig. (3.12). The internal forces can be written immediately as,

$$S = -Q_1, \qquad V = -Q_2, \qquad M = Q_2 x - Q_6, \qquad T = -Q_4. \tag{3.24}$$

Likewise, at a section x distance from node j, the internal forces are

$$S = -Q_7, \qquad V = -Q_8, \qquad M = -Q_8 x - Q_{12}, \qquad T = -Q_{10}. \tag{3.25}$$

Expanding Eq. (3.23) ($i = 1, 2, 4, 6$ and $i = 7, 8, 10, 12$) in accordance

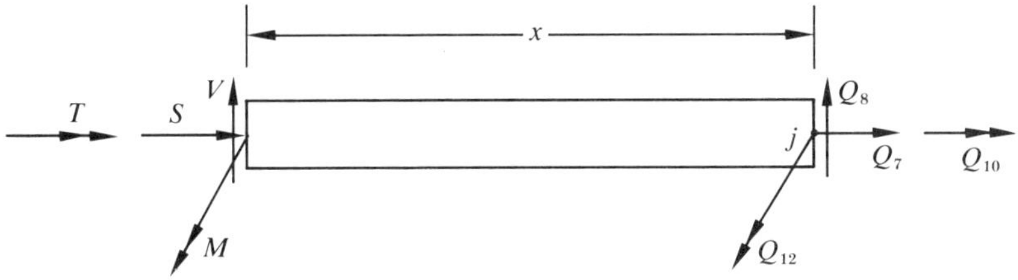

with Eqs. (3.24) and (3.25) respectively, and integrating yield,

$$q_1 = \eta_1 Q_1 \tag{a}$$

$$q_2 = -\eta_2 Q_6 + \eta_3 Q_2 \tag{b}$$

$$q_4 = \eta_5 Q_4 \tag{c}$$

$$q_6 = \eta_4 Q_6 - \eta_2 Q_2 \tag{d}$$

and

$$q_7 = \eta_1 Q_7 \tag{e}$$

$$q_8 = \eta_2 Q_{12} + \eta_3 Q_8 \tag{f}$$

$$q_{10} = \eta_5 Q_{10} \tag{g}$$

$$q_{12} = \eta_4 Q_{10} + \eta_2 Q_8 \tag{h}$$

where

$$\eta_1 = \frac{L}{AE}$$

$$\eta_2 = \frac{L^2}{2EI_z}$$

$$\eta_3 = \frac{L^3}{3EI_z} + \frac{L}{AG}$$

$$\eta_4 = \frac{L}{EI_z} \tag{3.26}$$

and

$$\eta_5 = \frac{L}{GJ}.$$

The two sets of Eqs. (a–d) and (e–h) give the force displacement relations at node i and j, respectively.

Equations (a–d) may be written in a matrix equation form as,

$$\begin{bmatrix} q_1 \\ q_2 \\ q_4 \\ q_6 \end{bmatrix} = \begin{bmatrix} \eta_1 & 0 & 0 & 0 \\ 0 & \eta_3 & 0 & \eta_2 \\ 0 & 0 & \eta_5 & 0 \\ 0 & \eta_2 & 0 & \eta_4 \end{bmatrix} \begin{bmatrix} Q_1 \\ Q_2 \\ Q_4 \\ Q_6 \end{bmatrix} \tag{i}$$

In our previous discussion it was stated that the stiffness coefficients of a structural element can be derived by successively giving each coordinate displacement a unit holding all others at zero and then calculating all the result coordinate forces. Thus, applying this principle to Eq. (i) yields the following four distinct conditions.

1. $q_1 = 1$ and all other q's are zero.
2. $q_2 = 1$ and all other q's are zero.
3. $q_4 = 1$ and all other q's are zero.
4. $q_5 = 1$ and all other q's are zero.

Figure 3.13 shows these conditions graphically.

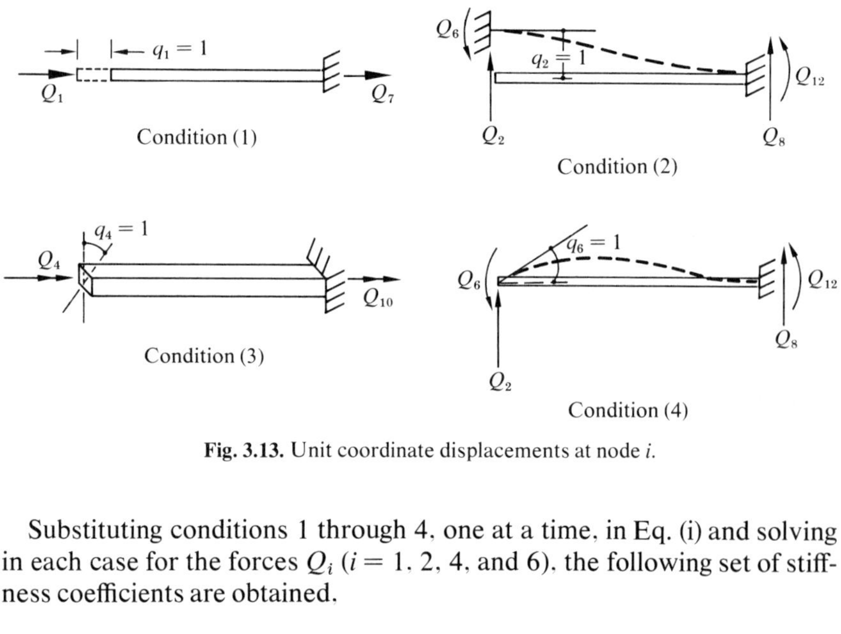

Fig. 3.13. Unit coordinate displacements at node *i*.

Substituting conditions 1 through 4, one at a time, in Eq. (i) and solving in each case for the forces Q_i ($i = 1, 2, 4$, and 6), the following set of stiffness coefficients are obtained.

$$\begin{aligned}
&\bar{S}_{11} = \alpha_1 \\
&\bar{S}_{21} = \bar{S}_{61} = \bar{S}_{41} = \bar{S}_{42} = \bar{S}_{64} = 0 \\
&\bar{S}_{22} = \alpha_4{}^z \\
&\bar{S}_{44} = \alpha_5 \\
&\bar{S}_{66} = \alpha_3{}^z \\
&\bar{S}_{62} = \alpha_2{}^z
\end{aligned} \tag{j}$$

where

$$\alpha_1 = \frac{1}{\eta_1}$$

$$\alpha_2^z = \frac{\eta_2}{\eta_3\eta_4 - \eta_2^2}$$

$$\alpha_3^z = \frac{\eta_3}{\eta_3\eta_4 - \eta_2^2} \qquad (3.27)$$

$$\alpha_4^z = \frac{\eta_4}{\eta_3\eta_4 - \eta_2^2}$$

and

$$\alpha_5 = \frac{1}{\eta_5}$$

$$\alpha_6^z = \frac{\eta_2 L - \eta_3}{\eta_3\eta_4 - \eta_2^2}$$

The superscript z which appears in Eq. (3.27) indicates that the moment of inertia of the element cross-section is taken about the z axis.

From equilibrium (*see* Fig. 3.12) the coordinate forces at node j resulting from unit coordinate displacements at node i are given by,

$$\begin{aligned} Q_7 &= -Q_1 \\ Q_8 &= -Q_2 \\ Q_{10} &= -Q_4 \\ Q_{12} &= -Q_6 + Q_2 L. \end{aligned} \qquad \text{(k)}$$

Recognizing that Q_i values ($i = 1, 2, 4,$ and 6) are equivalent to the stiffness coefficients for unit displacements, i.e.,

$$\frac{Q_1}{q_1} = \bar{S}_{11},$$

for instance, the following set of coefficients may be deduced by utilizing Eqs. (j) and (k),

$$\begin{aligned} \bar{S}_{71} &= -\alpha_1 \\ \bar{S}_{82} &= -\alpha_4^z \\ \bar{S}_{12,2} &= \alpha_2^z \\ \bar{S}_{10,4} &= -\alpha_5 \\ \bar{S}_{86} &= -\alpha_2^z \\ \bar{S}_{12,6} &= \alpha_6^z \end{aligned} \qquad \text{(l)}$$

and

$$\bar{S}_{81}=\bar{S}_{12,1}=\bar{S}_{10,1}=\bar{S}_{72}=\bar{S}_{10,2}=\bar{S}_{76}=\bar{S}_{10,6}=\bar{S}_{10,6}=\bar{S}_{74}=\bar{S}_{84}=\bar{S}_{12,4}=0.$$

To find the other coefficients which correspond to unit displacements at node j we follow the same procedure as we did at node i. Thus, rewriting Eqs. (e–h) in matrix form yields,

$$\begin{bmatrix} q_7 \\ q_8 \\ q_{10} \\ q_{12} \end{bmatrix} = \begin{bmatrix} \eta_1 & 0 & 0 & 0 \\ 0 & \eta_3 & 0 & \eta_2 \\ 0 & 0 & \eta_5 & 0 \\ 0 & \eta_2 & 0 & \eta_4 \end{bmatrix} \begin{bmatrix} Q_7 \\ Q_8 \\ Q_{10} \\ Q_{12} \end{bmatrix} \tag{m}$$

Here again we impose the following four conditions.

1. $q_7 = 1$ and all other q's are zero.
2. $q_8 = 1$ and all other q's are zero.
3. $q_{10} = 1$ and all other q's are zero.
4. $q_{12} = 1$ and all other q's are zero.

Graphically, these conditions are shown in Fig. 3.14.

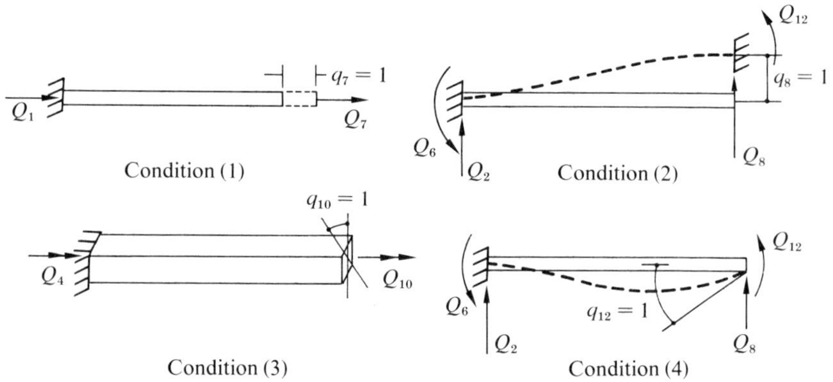

Fig. 3.14. Unit coordinate displacements at node j.

Substituting conditions 1 through 4, one at a time, in Eq. (m) and solving in each case for the forces Q_i ($i = 7, 8, 10$, and 12), the following set of stiffness coefficients is obtained.

$$\begin{aligned} \bar{S}_{77} &= \alpha_1 \\ \bar{S}_{88} &= \alpha_4{}^z \\ \bar{S}_{12,8} &= -\alpha_2{}^z \\ \bar{S}_{12,12} &= \alpha_3{}^z \\ \bar{S}_{10,10} &= \alpha_5 \end{aligned} \tag{n}$$

and

$$\bar{S}_{87} = \bar{S}_{12,7} = \bar{S}_{10,7} = \bar{S}_{10,8} = \bar{S}_{12,10} = 0.$$

The normal-to-plane stiffness coefficients associated with the coordinate displacements q_i ($i = 3, 5, 9$, and 11) can be derived in a similar manner as those for the in-plane case. But if we examine the forces Q_i ($i = 3, 5, 9$, and 11) closely, we can relate them to Q_i ($i = 2, 6, 8$, and 12), respectively, with the exception that the bending moments Q_i ($i = 5$ and 11) are in the opposite direction to the adopted sign convention. Therefore, performing the appropriate change in the signs of the stiffness coefficients associated with the unit displacements q_i ($i = 2, 6, 8$, and 12), the following coefficients corresponding to the unit displacements q_i ($i = 3, 5, 9$ and 11) are deduced.

$$\begin{aligned}
\bar{S}_{33} &= \bar{S}_{22}\ (Iy \text{ should replace } Iz) = \alpha_4{}^y \\
\bar{S}_{53} &= -\bar{S}_{62}\ (Iy \text{ should replace } Iz) = -\alpha_2{}^y \\
\bar{S}_{93} &= \bar{S}_{82}\ (Iy \text{ should replace } Iz) = \alpha_4{}^y \\
\bar{S}_{11,3} &= -\bar{S}_{12,2}\ (Iy \text{ should replace } Iz) = -\alpha_2{}^y \\
\bar{S}_{i3} &= 0\ (i = 4, 6, 7, 8, 10, \text{ and } 12) \\
\bar{S}_{55} &= \bar{S}_{66}\ (Iy \text{ should replace } Iz) = \alpha_3{}^y \\
\bar{S}_{95} &= \bar{S}_{86}\ (Iy \text{ should replace } Iz) = \alpha_2{}^y \\
\bar{S}_{11,5} &= \bar{S}_{12,6}\ (Iy \text{ should replace } Iz) = \alpha_6{}^y \\
\bar{S}_{i5} &= 0\ (i = 6, 7, 8, 10, \text{ and } 12) \\
\bar{S}_{99} &= \bar{S}_{ii}\ (Iy \text{ should replace } Iz) = \alpha_4{}^y \\
\bar{S}_{11,9} &= -\bar{S}_{12,8}\ (Iy \text{ should replace } Iz) = \alpha_2{}^y \\
\bar{S}_{i9} &= 0\ (i = 10 \text{ and } 12) \\
\bar{S}_{11,11} &= \bar{S}_{12,12}\ (Iy \text{ should replace } Iz) = \alpha_3{}^y \\
\bar{S}_{12,11} &= 0
\end{aligned} \tag{o}$$

With all the stiffness coefficients calculated (*see* Eqs. 3.22, j, p, n, and o) we can now write the following stiffness matrix equation relating forces and displacement for beam elements.

$$
\begin{bmatrix} F_i^{\bar{x}} \\ F_i^{\bar{y}} \\ F_i^{\bar{z}} \\ M_i^{\bar{x}} \\ M_i^{\bar{y}} \\ M_i^{\bar{z}} \\ F_j^{\bar{x}} \\ F_j^{\bar{y}} \\ F_j^{\bar{z}} \\ M_j^{\bar{x}} \\ M_j^{\bar{y}} \\ M_j^{\bar{z}} \end{bmatrix} = \begin{bmatrix}
\alpha_1 & & & & & & & & & & & \\
0 & \alpha_4{}^z & & & & & & & & & & \\
0 & 0 & \alpha_4{}^y & & & & & & & & & \\
0 & 0 & 0 & \alpha_5 & & & & & & & & \\
0 & 0 & -\alpha_2{}^y & 0 & \alpha_3{}^y & & & & & & & \\
0 & \alpha_2{}^z & 0 & 0 & 0 & \alpha_3{}^z & & & & & & \\
-\alpha_1 & 0 & 0 & 0 & 0 & 0 & \alpha_1 & & & & & \\
0 & -\alpha_4{}^z & 0 & 0 & 0 & -\alpha_2{}^z & 0 & \alpha_4{}^z & & & & \\
0 & 0 & -\alpha_4{}^y & 0 & \alpha_2{}^y & 0 & 0 & 0 & \alpha_4{}^y & & & \\
0 & 0 & 0 & -\alpha_5 & 0 & 0 & 0 & 0 & 0 & \alpha_5 & & \\
0 & 0 & -\alpha_2{}^y & 0 & \alpha_6{}^y & 0 & 0 & 0 & \alpha_2{}^y & 0 & \alpha_3{}^y & \\
0 & \alpha_2{}^z & 0 & 0 & 0 & \alpha_6{}^z & 0 & -\alpha_2{}^z & 0 & 0 & 0 & \alpha_3{}^z
\end{bmatrix} \begin{bmatrix} \delta_i^{\bar{x}} \\ \delta_i^{\bar{y}} \\ \delta_i^{\bar{z}} \\ \theta_i^{\bar{x}} \\ \theta_i^{\bar{y}} \\ \theta_i^{\bar{z}} \\ \delta_j^{\bar{x}} \\ \delta_j^{\bar{y}} \\ \delta_j^{\bar{z}} \\ \theta_j^{\bar{x}} \\ \theta_j^{\bar{y}} \\ \theta_j^{\bar{z}} \end{bmatrix} \tag{3.28}
$$

where the α's are defined in Eqs. (3.27) and (3.26).

Equation (3.28) gives the functional relationship between coordinate forces and displacements in a local coordinate system.

For beam elements which are arbitrarily oriented, the transformation given in Eq. (3.13) should be performed. Thus, in the reference coordinate system, Eq. (3.28) takes on the following form,

$$
\begin{bmatrix} F_i^x \\ F_i^y \\ F_i^z \\ M_i^x \\ M_i^y \\ M_i^z \\ F_j^x \\ F_j^y \\ F_j^z \\ M_j^x \\ M_j^y \\ M_j^z \end{bmatrix} = [S] \begin{bmatrix} \delta_i^x \\ \delta_i^y \\ \delta_i^z \\ \theta_i^x \\ \theta_i^y \\ \theta_i^z \\ \delta_j^x \\ \delta_j^y \\ \delta_j^z \\ \theta_j^x \\ \theta_j^y \\ \theta_j^z \end{bmatrix} \tag{3.29}
$$

Where $[S]$ is given by,

$$
[S] = \begin{bmatrix} [T] & 0 & 0 & 0 \\ 0 & [T] & 0 & 0 \\ 0 & 0 & [T] & 0 \\ 0 & 0 & 0 & [T] \end{bmatrix}^{-1} \times [\bar{S}] \times \begin{bmatrix} [T] & 0 & 0 & 0 \\ 0 & [T] & 0 & 0 \\ 0 & 0 & [T] & 0 \\ 0 & 0 & 0 & [T] \end{bmatrix} \tag{3.30}
$$

The transformation matrix $[T]$ is given in Eq. (3.15) and $[\bar{S}]$ is in Eq. (3.28). If the matrix multiplication is carried out in Eq. (3.30) it will yield the following generalized stiffness coefficients for a beam element having a spatial orientation in space,

$$
\begin{aligned}
S_{11} &= T_{11}^2\alpha_1 + T_{12}^2\alpha_4{}^z + T_{13}^2\alpha_4{}^y \\
S_{12} &= T_{11}T_{21}\alpha_1 + T_{12}T_{22}\alpha_4{}^z + T_{13}T_{23}\alpha_4{}^y \\
S_{13} &= T_{11}T_{31}\alpha_1 + T_{12}T_{32}\alpha_4{}^z + T_{13}T_{33}\alpha_4{}^y \\
S_{14} &= -T_{12}T_{13}\alpha_2{}^y + T_{12}T_{13}\alpha_2{}^z \\
S_{15} &= -T_{12}T_{23}\alpha_2{}^y + T_{13}T_{22}\alpha_2{}^z \\
S_{16} &= -T_{12}T_{33}\alpha_2{}^y + T_{13}T_{32}\alpha_2{}^z \\
S_{17} &= -S_{11} \\
S_{18} &= -S_{12} \\
S_{19} &= -S_{13} \\
S_{1,10} &= S_{14} \\
S_{1,11} &= S_{15} \\
S_{1,12} &= S_{16} \\
S_{21} &= S_{12} \\
S_{22} &= T_{21}^2\alpha_1 + T_{22}^2\alpha_4{}^z + T_{23}^2\alpha_4{}^y \qquad (3.30a) \\
S_{23} &= T_{21}T_{31}\alpha_1 + T_{22}T_{32}\alpha_4{}^z + T_{23}T_{33}\alpha_4{}^y \\
S_{24} &= -T_{13}T_{22}\alpha_2{}^y + T_{12}T_{23}\alpha_2{}^z \\
S_{25} &= -T_{22}T_{23}\alpha_2{}^y + T_{22}T_{23}\alpha_2{}^z \\
S_{26} &= -T_{22}T_{33}\alpha_2{}^y + T_{23}T_{32}\alpha_2{}^z \\
S_{27} &= -S_{12} \\
S_{28} &= -S_{22} \\
S_{29} &= -S_{23} \\
S_{2,10} &= S_{24} \\
S_{2,11} &= S_{25} \\
S_{2,12} &= S_{26} \\
S_{3i} &= S_{i3} \qquad (i = 1, 2) \\
S_{33} &= T_{31}^2\alpha_1 + T_{32}^2\alpha_4{}^z + T_{33}^2\alpha_4{}^y \\
S_{34} &= -T_{13}T_{32}\alpha_2{}^y + T_{12}T_{33}\alpha_2{}^z
\end{aligned}
$$

$$S_{35} = -T_{23}T_{32}\alpha_2{}^y + T_{22}T_{33}\alpha_2{}^z$$
$$S_{36} = -T_{32}T_{33}\alpha_2{}^y + T_{32}T_{33}\alpha_2{}^z$$
$$S_{37} = -S_{13}$$
$$S_{38} = -S_{23}$$
$$S_{39} = -S_{33}$$
$$S_{3,10} = S_{34}$$
$$S_{3,11} = S_{35}$$
$$S_{3,12} = S_{36}$$
$$S_{4i} = S_{i4} \qquad (i = 1, 2, 3)$$
$$S_{44} = T_{11}^2\alpha_5 + T_{12}^2\alpha_3{}^y + T_{13}^2\alpha_3{}^z$$
$$S_{45} = T_{11}T_{21}\alpha_5 + T_{12}T_{22}\alpha_3{}^y + T_{13}T_{23}\alpha_3{}^z$$
$$S_{46} = T_{11}T_{31}\alpha_5 + T_{12}T_{32}\alpha_3{}^y + T_{13}T_{33}\alpha_3{}^z$$
$$S_{47} = -S_{14}$$
$$S_{48} = -S_{15} \tag{3.30a}$$
$$S_{49} = -S_{16}$$
$$S_{4,10} = -T_{11}^2\alpha_5 + T_{12}^2\alpha_6{}^y + T_{13}^2\alpha_6{}^z$$
$$S_{4,11} = -T_{11}T_{21}\alpha_5 + T_{12}T_{22}\alpha_6{}^y + T_{13}T_{23}\alpha_6{}^z$$
$$S_{4,12} = -T_{11}T_{31}\alpha_5 + T_{12}T_{32}\alpha_6{}^y + T_{13}T_{33}\alpha_6{}^z$$
$$S_{5i} = S_{i5} \qquad (i = 1, 2, 3, 4)$$
$$S_{55} = T_{21}^2\alpha_5 + T_{22}^2\alpha_3{}^y + T_{23}^2\alpha_3{}^z$$
$$S_{56} = T_{21}T_{31}\alpha_5 + T_{22}T_{32}\alpha_3{}^y + T_{23}T_{33}\alpha_3{}^z$$
$$S_{57} = -S_{24}$$
$$S_{58} = -S_{25}$$
$$S_{59} = -S_{26}$$
$$S_{5,10} = -T_{11}T_{21}\alpha_5 + T_{12}T_{22}\alpha_6{}^y + T_{13}T_{23}\alpha_6{}^z$$
$$S_{5,11} = -T_{21}^2\alpha_5 + T_{22}^2\alpha_6{}^y + T_{23}^2\alpha_6{}^z$$
$$S_{5,12} = -T_{21}T_{31}\alpha_5 + T_{22}T_{32}\alpha_6{}^y + T_{23}T_{33}\alpha_6{}^z$$
$$S_{6i} = S_{i6} \qquad (i = 1, 2, \ldots, 5)$$

$$
\begin{aligned}
S_{66} &= T_{31}^2\alpha_5 + T_{32}^2\alpha_3{}^y + T_{33}^2\alpha_3{}^z \\
S_{67} &= S_{34} \\
S_{68} &= S_{35} \\
S_{69} &= -S_{36} \\
S_{6,10} &= -T_{11}T_{31}\alpha_5 + T_{12}T_{32}\alpha_6{}^y + T_{13}T_{33}\alpha_6{}^z \\
S_{6,11} &= -T_{21}T_{31}\alpha_5 + T_{22}T_{32}\alpha_6{}^y + T_{23}T_{33}\alpha_6{}^z \\
S_{6,12} &= -T_{31}^2\alpha_5 + T_{32}^2\alpha_6{}^y + T_{33}^2\alpha_6{}^z \\
S_{7i} &= S_{i7} \qquad (i = 1, 2, \ldots, 6) \\
S_{77} &= S_{11} \\
S_{78} &= S_{12} \\
S_{79} &= S_{13} \\
S_{7,10} &= -T_{12}T_{13}\alpha_2{}^z + T_{12}T_{13}\alpha_2{}^y \\
S_{7,11} &= -T_{12}T_{23}\alpha_2{}^z + T_{13}T_{22}\alpha_2{}^y \\
S_{7,12} &= -T_{12}T_{33}\alpha_2{}^z + T_{13}T_{32}\alpha_2{}^y \\
S_{8i} &= S_{i8} \qquad (i = 1, 2, \ldots, 7) \\
S_{88} &= S_{22} \\
S_{89} &= S_{23} \\
S_{8,10} &= -T_{13}T_{22}\alpha_2{}^z + T_{12}T_{23}\alpha_2{}^y \\
S_{8,11} &= -T_{22}T_{23}\alpha_2{}^z + T_{22}T_{23}\alpha_2{}^y \\
S_{8,12} &= -T_{22}T_{33}\alpha_2{}^z + T_{23}T_{32}\alpha_2{}^y \\
S_{9i} &= S_{i9} \qquad (i = 1, 2, \ldots, 8) \\
S_{99} &= S_{33} \\
S_{9,10} &= -T_{13}T_{32}\alpha_2{}^z + T_{12}T_{33}\alpha_2{}^y \\
S_{9,11} &= -T_{23}T_{32}\alpha_2{}^z + T_{22}T_{33}\alpha_2{}^y \\
S_{9,12} &= -T_{32}T_{33}\alpha_2{}^z + T_{32}T_{33}\alpha_2{}^y \\
S_{10,i} &= S_{i,10} \qquad (i = 1, 2, \ldots, 9) \\
S_{10,10} &= S_{44} \\
S_{10,11} &= S_{45}
\end{aligned}
\tag{3.30a}
$$

$$
\begin{aligned}
S_{10,12} &= S_{46} \\
S_{11,i} &= S_{i,11} \qquad (i = 1, 2, \ldots, 10) \\
S_{11,11} &= S_{55} \\
S_{11,12} &= S_{56} \\
S_{12,i} &= S_{i,12} \qquad (i = 1, 2, \ldots, 11) \\
S_{12,12} &= S_{66}
\end{aligned}
\tag{3.30a}
$$

3.10 SPECIAL CASES FOR THE RECTILINEAR BEAM ELEMENT

Case 1. Three-dimensional beam element with transverse shear deformation neglected.

If transverse shear deformation is neglected, which is the case in the narrow slender beams, then the stiffness coefficients in Eq. (3.28) take on the following values of α's.

$$
\begin{aligned}
\alpha_1 &= \frac{AE}{L} \\
\alpha_2^{z}, \alpha_2^{y} &= \frac{6EI_z}{L^2}, \frac{6EI_y}{L^2} \\
\alpha_3^{z}, \alpha_3^{y} &= \frac{4EI_z}{L}, \frac{4EI_y}{L} \\
\alpha_4^{z}, \alpha_4^{y} &= \frac{12EI_z}{L^3}, \frac{12EI_y}{L^3} \\
\alpha_5 &= \frac{GJ}{L} \\
\alpha_6^{z}, \alpha_6^{y} &= \frac{2EI_z}{L}, \frac{2EI_y}{L}
\end{aligned}
\tag{3.31}
$$

Case 2. Two-dimensional beam elements.

In the case of two-dimensional beam elements (Fig. 3.15), the stiffness matrix Eq. (3.28) reduces to,

$$
\begin{bmatrix} F_i^{\bar{x}} \\ F_i^{\bar{y}} \\ M_i^{\bar{z}} \\ F_j^{\bar{x}} \\ F_j^{\bar{y}} \\ M_j^{\bar{z}} \end{bmatrix}
=
\begin{bmatrix}
\alpha_1 & & & & & \\
0 & \alpha_4^{z} & & \text{Sym.} & & \\
0 & \alpha_2^{z} & \alpha_3^{z} & & & \\
-\alpha_1 & 0 & 0 & \alpha_1 & & \\
0 & -\alpha_4^{z} & -\alpha_2^{z} & 0 & \alpha_4^{z} & \\
0 & \alpha_2^{z} & \alpha_6^{z} & 0 & -\alpha_2^{z} & \alpha_3^{z}
\end{bmatrix}
\begin{bmatrix} \delta_i^{\bar{x}} \\ \delta_i^{\bar{y}} \\ \theta_i^{\bar{z}} \\ \delta_j^{\bar{x}} \\ \delta_j^{\bar{y}} \\ \theta_j^{\bar{z}} \end{bmatrix}
\tag{3.32}
$$

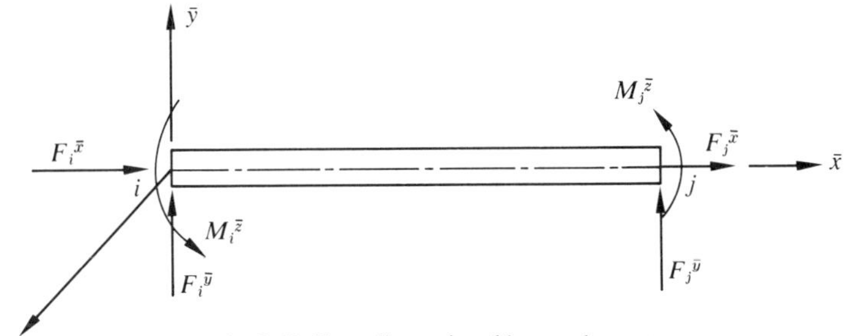

Fig. 3.15. Two-dimensional beam element.

where the α's are given in Eqs. (3.27) and (3.26). Use Eq. (3.31) for α's if shear deformation is negligible.

In reference coordinate system the stiffness matrix in Eq. (3.32) becomes,

$$[S] = \begin{bmatrix} [T] & 0 \\ 0 & [T] \end{bmatrix}^{1} [S] \begin{bmatrix} [T] & 0 \\ 0 & [T] \end{bmatrix} \tag{3.33}$$

Therefore, Eq. (3.32) in reference coordinate system can be written as,

$$\begin{bmatrix} F_i^x \\ F_i^y \\ M_i^z \\ F_j^x \\ F_j^y \\ M_j^z \end{bmatrix} = [S] \begin{bmatrix} \delta_i^x \\ \delta_i^y \\ \theta_i^z \\ \delta_j^x \\ \delta_j^y \\ \theta_j^z \end{bmatrix} \tag{3.34}$$

where $[S]$ can be deduced from Eq. (3.30a) and is given by,

$$[S] = \begin{bmatrix}
\begin{pmatrix} T_{11}^2\alpha_1 \\ + \\ T_{21}^2\alpha_4^z \end{pmatrix} & & & & & \\
\begin{pmatrix} T_{11}T_{12}\alpha_1 \\ + \\ T_{21}T_{22}\alpha_4^z \end{pmatrix} & \begin{pmatrix} T_{12}^2\alpha_1 \\ + \\ T_{22}^2\alpha_4^z \end{pmatrix} & & \text{Sym.} & & \\
(T_{21}\alpha_2^z) & (T_{22}\alpha_2^z) & (\alpha_3^z) & & & \\
\begin{pmatrix} -T_{21}^2\alpha_4^z \\ - \\ T_{11}^2\alpha_1 \end{pmatrix} & \begin{pmatrix} -T_{21}T_{22}\alpha_4^z \\ - \\ T_{12}T_{11}\alpha_1 \end{pmatrix} & (-T_{21}\alpha_2^z) & \begin{pmatrix} T_{11}^2\alpha_1 \\ + \\ T_{21}^2\alpha_4^z \end{pmatrix} & & \\
\begin{pmatrix} -T_{22}T_{21}\alpha_4^z \\ - \\ T_{11}T_{12}\alpha_1 \end{pmatrix} & \begin{pmatrix} -T_{12}^2\alpha_1 \\ - \\ T_{22}^2\alpha_4^z \end{pmatrix} & (-T_{22}\alpha_2^z) & \begin{pmatrix} T_{12}T_{11}\alpha_1 \\ + \\ T_{22}T_{21}\alpha_4^z \end{pmatrix} & \begin{pmatrix} T_{12}^2\alpha_1 \\ + \\ T_{22}^2\alpha_4^z \end{pmatrix} & \\
(T_{21}\alpha_2^z) & (T_{22}\alpha_2^z) & (\alpha_6^z) & (-T_{21}\alpha_2^z) & (-T_{22}\alpha_2^z) & (\alpha_3^z)
\end{bmatrix}$$

(3.34a)

Illustrative Examples

Example 7

Find the internal forces and node deflections of the rigid frame structure shown below. Neglect shear deformations. Assume $EI = 15 \times 10^6$ lb-in^2 and $AE = 45 \times 10^6$ lbs.

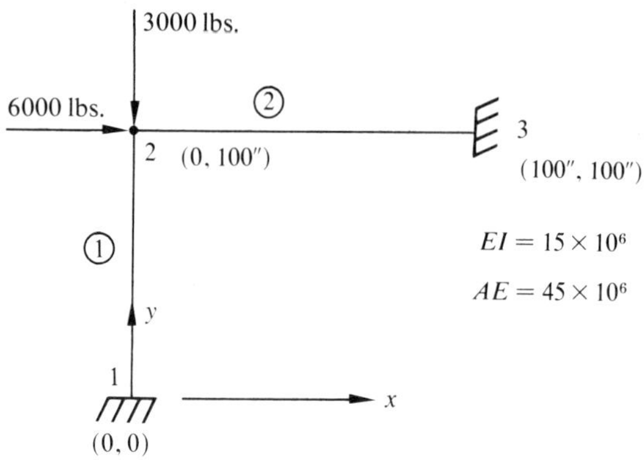

Node Coordinates				Beam Length	
Node	x	y		Beam	Length
1	0	0		1	100
2	0	100		2	100
3	100	100			

Coefficients of Transformation Matrix (Eq. 3.16).

Beam 1

$$T_{11} = 0, \qquad T_{12} = 1, \qquad T_{21} = -1, \qquad T_{22} = 0$$

Beam 2

$$T_{11} = 1, \qquad T_{12} = 0, \qquad T_{21} = 0, \qquad T_{22} = 1$$

The individual element stiffness matrices from Eq. (3.34a) are

Beam 1

$$[S]_1 = 10^4 \begin{bmatrix} 0.018 & & & & & \\ 0 & 45 & & & \text{Sym.} & \\ -0.9 & 0 & 60 & & & \\ -0.018 & 0 & 0.9 & 0.018 & & \\ 0 & -45 & 0 & 0 & 45 & \\ -0.9 & 0 & 30 & 0.9 & 0 & 60 \end{bmatrix}$$

Beam 2

$$[S]_2 = 10^4 \begin{bmatrix} 45 & & & & & \\ 0 & 0.018 & & \text{Sym.} & & \\ 0 & 0.9 & 60 & & & \\ -45 & 0 & 0 & 45 & & \\ 0 & -0.018 & -0.9 & 0 & 0.018 & \\ 0 & 0.9 & 30 & 0 & -0.9 & 60 \end{bmatrix}$$

Boundary conditions: $\delta_1^x = \delta_1^y = \theta_1^z = \delta_3^x = \delta_3^y = \theta_3^z = 0$
Applied Forces: $F_2^x = 6000\#$, $F_2^y = -3000\#$

The aggregate stiffness matrix is,

$$[S] = [S]_1 + [S]_2 = 10^4 \begin{bmatrix} 0.018 & 0 & -0.9 & -0.018 & 0 & -0.9 & & & \\ 0 & 45 & 0 & 0 & -45 & 0 & & & \\ -0.9 & 0 & 60 & 0.9 & 0 & 30 & & & \\ -0.018 & 0 & 0.9 & 0.018+45 & 0+0 & 0.9+0 & -45 & 0 & 0 \\ 0 & -45 & 0 & 0+0 & 45+0.018 & 0+0.9 & 0 & -0.018 & 0.9 \\ -0.9 & 0 & 30 & 0.9+0 & 0+0.9 & 60+60 & 0 & -0.9 & 30 \\ & & & -45 & 0 & 0 & 45 & 0 & 0 \\ & & & 0 & -0.018 & -0.9 & 0 & 0.018 & -0.9 \\ & & & 0 & 0.9 & 30 & 0 & -0.9 & 60 \end{bmatrix}$$

Therefore, the reduced aggregate stiffness matrix equation is,

$$\begin{bmatrix} 60^K \\ -30^K \\ 0 \end{bmatrix} = 10^4 \begin{bmatrix} 45.018 & 0 & 0.9 \\ 0 & 45.018 & 0.9 \\ 0.9 & 0.9 & 120 \end{bmatrix} \begin{bmatrix} \delta_2^x \\ \delta_2^y \\ \theta_2^z \end{bmatrix}$$

Solving for the displacements yields,

$$\delta_2^x = 0.133 \text{ in.}$$

$$\delta_2^y = -0.066 \text{ in.}$$

$$\theta_2^z = -0.000489 \text{ rad.}$$

Internal loads from individual element stiffness matrix equations are

Beam 1

$$F_1^x = -21.6 \text{ lbs}$$

$$F_1^y = 2998.5 \text{ lbs}$$

$$M_1^z = 1054 \text{ in.-lbs}$$

$$F_2^x = 21.6 \text{ lbs}$$

$$F_2^y = -2998.5 \text{ lbs}$$

$$M_2^z = 1108 \text{ in.-lbs}$$

Beam 2

$$F_2^x = 5999 \text{ lbs}$$
$$F_2^y = -4.5 \text{ lbs}$$
$$M_2^z = -1475 \text{ in.-lbs}$$
$$F_3^x = -5999 \text{ lbs}$$
$$F_3^y = 4.5 \text{ lbs}$$
$$M_3^z = -1035 \text{ in.-lbs}$$

Example 8

In the beam shown below, find the displacements at the point of application of the load P. Assume EI and AG to be constant and neglect axial deformation.

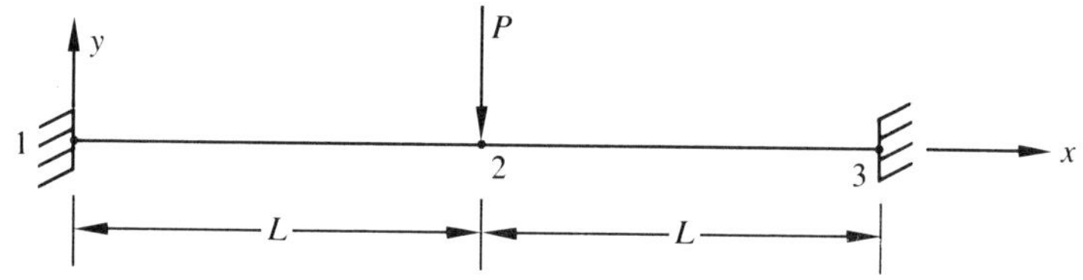

Node Coordinates		
Node	x	y
1	0	0
2	L	0
3	$2L$	0

Beam Length	
Beam	Length
1	L
2	L

Coefficients of transformation matrix,

Beams 1 and 2

$$T_{11} = 1, \qquad T_{12} = 0, \qquad T_{21} = 0, \qquad T_{22} = 1$$

Individual beam element stiffness matrix equations,

Beam 1

$$\begin{bmatrix} F_1^x \\ F_1^y \\ M_1^z \\ F_2^x \\ F_2^y \\ M_2^z \end{bmatrix} = \begin{bmatrix} 0 & 0 & 0 & 0 & 0 & 0 \\ 0 & \alpha_4^z & \alpha_2^z & 0 & -\alpha_4^z & \alpha_2^z \\ 0 & \alpha_2^z & \alpha_3^z & 0 & -\alpha_2^z & \alpha_6^z \\ 0 & 0 & 0 & 0 & 0 & 0 \\ 0 & -\alpha_4^z & -\alpha_2^z & 0 & \alpha_4^z & -\alpha_2^z \\ 0 & \alpha_2^z & \alpha_6^z & 0 & -\alpha_2^z & \alpha_3^z \end{bmatrix} \begin{bmatrix} \delta_1^x \\ \delta_1^y \\ \theta_1^z \\ \delta_2^x \\ \delta_2^y \\ \theta_2^z \end{bmatrix} \tag{a}$$

Beam 2

$$\begin{bmatrix} F_2^x \\ F_2^y \\ M_2^z \\ F_3^x \\ F_3^y \\ M_3^z \end{bmatrix} = \begin{bmatrix} 0 & 0 & 0 & 0 & 0 & 0 \\ 0 & \alpha_4^z & \alpha_2^z & 0 & -\alpha_4^z & \alpha_2^z \\ 0 & \alpha_2^z & \alpha_3^z & 0 & -\alpha_2^z & \alpha_6^z \\ 0 & 0 & 0 & 0 & 0 & 0 \\ 0 & -\alpha_4^z & -\alpha_2^z & 0 & \alpha_4^z & -\alpha_2^z \\ 0 & \alpha_2^z & \alpha_6^z & 0 & -\alpha_2^z & \alpha_3^z \end{bmatrix} \begin{bmatrix} \delta_2^x \\ \delta_2^y \\ \theta_2^z \\ \delta_3^x \\ \delta_3^y \\ \theta_3^z \end{bmatrix} \qquad \text{(b)}$$

Superposing Eqs. (a) and (b) and noticing the zero rows and columns yield the following aggregate stiffness matrix equation,

$$\begin{bmatrix} F_1^y \\ M_1^z \\ -P \\ 0 \\ F_3^y \\ M_3^z \end{bmatrix} = \begin{bmatrix} \alpha_4 & \alpha_2 & -\alpha_4 & \alpha_2 & & \\ \alpha_2 & \alpha_3 & -\alpha_2 & \alpha_6 & & \\ -\alpha_5 & -\alpha_2 & 2\alpha_4 & 0 & -\alpha_4 & \alpha_2 \\ \alpha_2 & \alpha_6 & 0 & 2\alpha_3 & -\alpha_2 & \alpha_6 \\ & & -\alpha_4 & -\alpha_2 & \alpha_4 & -\alpha_2 \\ & & \alpha_2 & \alpha_6 & -\alpha_2 & \alpha_3 \end{bmatrix} \begin{bmatrix} 0 \\ 0 \\ \delta_2^y \\ 0 \\ 0 \\ 0 \end{bmatrix} \qquad \text{(c)}$$

Note that the superscript z in Eq. (c) has been left out for brevity.

Deleting rows and corresponding columns where zero displacements appear in Eq. (c), yield,

$$-P = 2\alpha_4\delta_2 \qquad \text{or} \qquad \delta_2^y = -P/2\alpha_4,$$

where $\alpha_4 = 1/(L^3/12EI + L/AG)$ (shear deformation is included) or

$\alpha_4 = 12EI/L^3$ (shear deformation is neglected).

Example 9

In the rigid frame structure shown, find all displacements and internal forces. Use stiffness matrix method. Assume $E = 30 \times 10^6$ psi, $I = 2$ in^4, and $A = 6$ in^2 for all beam elements.

Writing the individual element matrix for each beam, superposing, and applying boundary conditions, yield the following reduced aggregate matrix equation for the structure.

$$\begin{bmatrix} 2\times 10^3 \\ -2.16\times 10^3 \\ -1.3\times 10^5 \\ 0 \\ -2.16\times 10^3 \\ 1.3\times 10^5 \end{bmatrix} = 10^4 \begin{bmatrix} 50 & 0 & 0.63 & -50 & 0 & 0 \\ 0 & 75 & 0.28 & 0 & -0.002 & 0.28 \\ 0.63 & 0.28 & 167 & 0 & -0.28 & 33 \\ -50 & 0 & 0 & 50 & 0 & 0.63 \\ 0 & -0.002 & -0.28 & 0 & 75 & -0.28 \\ 0 & 0.28 & 33 & 0.63 & -0.28 & 167 \end{bmatrix} \begin{bmatrix} \delta_2^x \\ \delta_2^y \\ \theta_2^z \\ \delta_3^x \\ \delta_3^y \\ \theta_3^z \end{bmatrix}$$

Inverting the stiffness matrix and solving for the displacements yield,

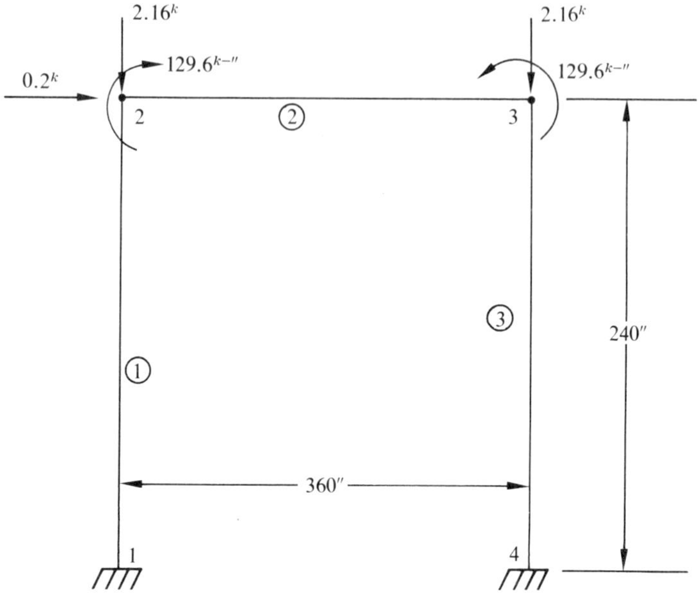

$$\delta_2^x = 3.07 \text{ in.} \qquad \delta_3^x = 3.07 \text{ in.}$$
$$\delta_2^y = -2.8 \times 10^{-3} \text{ in.} \qquad \delta_3^y = -2.95 \times 10^{-3}$$
$$\theta_2^z = -0.11 \text{ rad} \qquad \theta_3^z = -0.088 \text{ rad}$$

With the displacements now known, the internal forces can be obtained from the individual element matrices and are shown on the following sketch below.

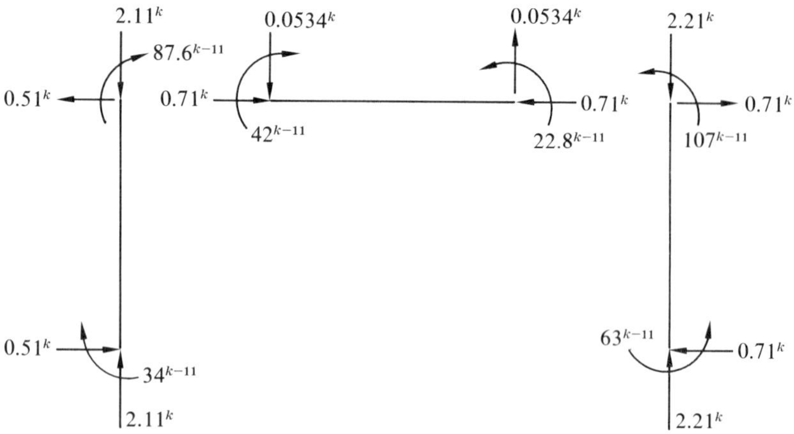

3.11 CURVED BEAM ELEMENT–STIFFNESS MATRIX

Utilization of straight structural members in the design of structural systems is not always possible. Structural efficiency as well as the shape configuration plays a major role in the selection for the structural constituents in a system. In aerospace for instance, curved structural members are widely used in atmospheric and space vehicles. In sea-based vehicles, such as submarines, the hulls are a common place for curved structural members. In civil engineering also, curved structural members are frequently utilized in bridges, buildings, and other structures where appearance and structural efficiency are of importance.

Therefore, it is worthwhile but not an extreme necessity, to develop the stiffness matrix equation for a curved beam element. The fact that a curved beam such as shown in Fig. 3.16(a) can always be idealized into a finite number of straight beam elements as in Fig. 3.16(b), justifies that the development is not extremely important. On the other hand, if we notice the increase in the number of unknowns which are introduced in the idealized case over what is in the unidealized condition, it makes it worth the effect to develop the curved beam stiffness matrix. On the part of accuracy, if enough segments are taken in case (b), very close results will be attained in comparison with the exact solution of the original structure.

Figure 3.17 shows a curved beam element of arbitrary arc length,

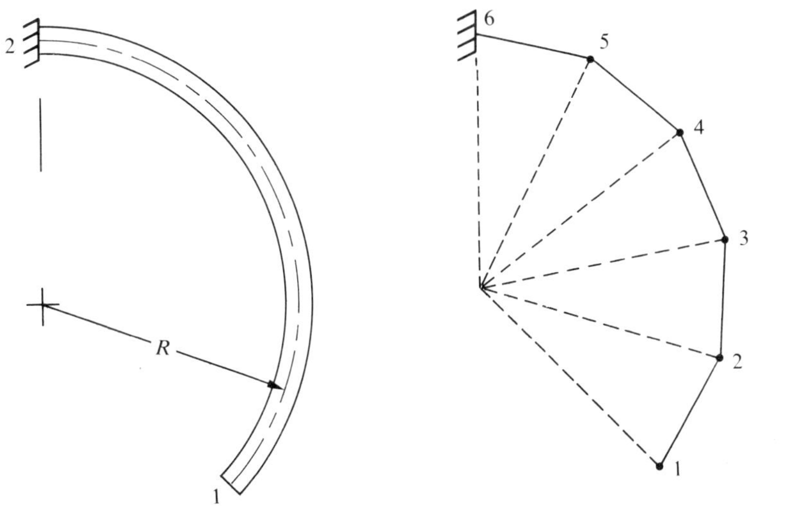

(a) Original Structure (b) Idealized into 5 straight segments

Fig. 3.16. Curved beam.

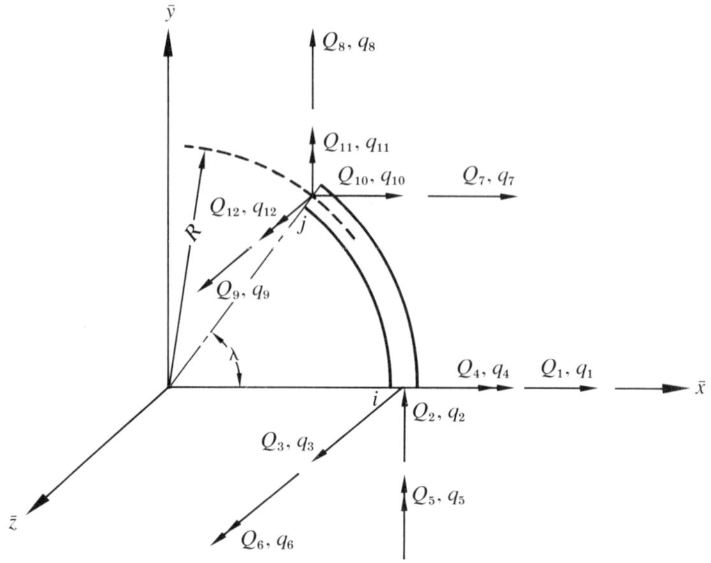

Fig. 3.17. Curved beam element.

radius, and of constant cross-section. Six coordinate forces and displacements are assumed to exist at each node and are positive as shown. The angle λ is the angle included between the lines suspended from nodes i and j. For convenience, the generalized force and displacement notation will be used while in the process of deriving the stiffness coefficients. This notation is given in Eqs. (2.1) and (2.2).

Since there are twelve distinct displacements (six degrees of freedom at each node), the order of the stiffness matrix will be 12×12. The stiffness coefficients for the curved beam element will be derived in the same manner as that of the straight beam. Further, it can be shown that there is no static coupling between the displacements in the $\bar{x}\bar{y}$ plane and those in the $\bar{x}\bar{z}$ plane.

Thus, we can derive the stiffness matrix in two separate groups:

1. Considering only displacements in $\bar{x}\bar{y}$ plane.
2. Considering only displacements in $\bar{x}\bar{z}$ plane.

In general, the strain energy can be expressed as the sum of the energies due to axial, bending, transverse shear and torsional forces, respectively, as follows,

$$U = \tfrac{1}{2} \int_0^{\lambda} \left\{ \frac{[S(\theta)]^2}{AE} + \frac{[M(\theta)]^2}{EI} + \frac{[+(\theta)]^2}{AG} + \frac{[T(\theta)]^2}{GJ} \right\} R\,d\theta. \qquad (3.35)$$

Performing partial differentiation on Eq. (3.35) with respect to a force Q_i, will yield a displacement q_i in the direction of Q_i, or,

$$q_i = \frac{\partial U}{\partial Q_i} = \int_0^\lambda \left\{ \frac{S}{AE}\frac{\partial S}{\partial Q_i} + \frac{M}{EI}\frac{\partial M}{\partial Q_i} + \frac{V}{AG}\frac{\partial V}{\partial Q_i} + \frac{T}{GJ}\frac{\partial T}{\partial Q_i} \right\} R\,d\theta. \quad (3.36)$$

On the basis of the assumption that displacements which are confined to the $\bar{x}\bar{y}$ plane (in this case they are q_1, q_2, q_6, q_7, q_8, and q_{12}) induce no forces in $\bar{x}\bar{z}$ plane ($Q_3 = Q_4 = Q_5 = Q_9 = Q_{10} = Q_{11} = 0$) we can deduce the following stiffness coefficients are zero,

$$S_{3i} = S_{4i} = S_{10,i} = S_{5i} = S_{9i} = S_{11,i} = 0 \qquad (i = 1, 2, 6, 7, 8, \text{and } 12) \quad (3.37)$$

To find the other stiffness coefficients, let us express the internal forces in Eq. (3.36) in terms of the coordinate forces at node i. These forces can be easily obtained from Fig. 3.18 and are given as follows:

$$\begin{aligned} S(\theta) &= Q_1 \sin\theta - Q_2 \cos\theta \\ V(\theta) &= Q_1 \cos\theta + Q_2 \sin\theta \\ M(\theta) &= -Q_6 - Q_1 R \sin\theta - Q_2 R(1 - \cos\theta) \end{aligned} \quad (3.38)$$

Utilizing Eq. (3.38) in Eq. (3.36), integrating and expanding for Q_i $(i = 1, 2, 6)$ yields,

$$\begin{bmatrix} q_1 \\ q_2 \\ q_6 \end{bmatrix} = \begin{bmatrix} \psi_1^z & \psi_2^z & \psi_3^z \\ \psi_2^z & \psi_4^z & \psi_5^z \\ \psi_3^z & \psi_5^z & \psi_6^z \end{bmatrix} \begin{bmatrix} Q_1 \\ Q_2 \\ Q_3 \end{bmatrix} \quad \text{(a)}$$

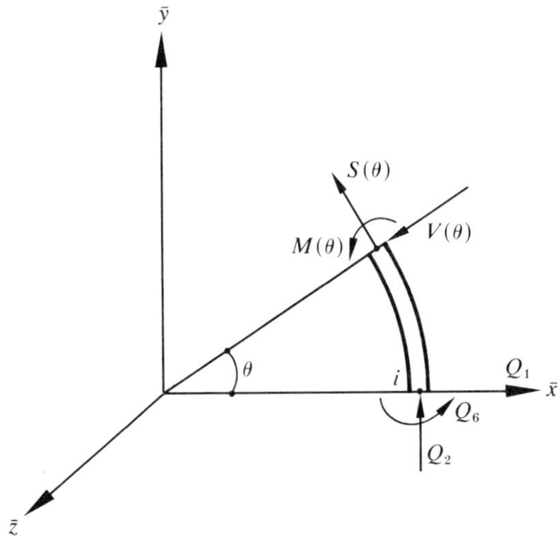

Fig. 3.18. Section of a curved beam element.

where

$$\psi_1{}^z = \left(\frac{R}{AE}+\frac{R^3}{EI_z}\right)\left(\frac{\lambda}{2}-\frac{\sin 2\lambda}{4}\right)+\frac{R}{AG}\left(\frac{\lambda}{2}+\frac{\sin 2\lambda}{4}\right)$$

$$\psi_2{}^z = -\left(\frac{R}{AE}+\frac{R^3}{EI_z}-\frac{R}{AG}\right)\frac{\sin^2\lambda}{2}+\frac{R^3}{EI_z}(1-\cos\lambda)$$

$$\psi_3{}^z = \frac{R^3}{EI_z}(1-\cos\lambda)$$

$$\psi_4{}^z = \left(\frac{R}{AE}+\frac{R^3}{EI_z}\right)\left(\frac{\lambda}{2}+\frac{\sin 2\lambda}{4}\right)+\frac{R^3}{EI_z}(\lambda-2\sin\lambda)+\frac{R}{AG}\left(\frac{\lambda}{2}-\frac{\sin 2\lambda}{4}\right) \tag{3.39}$$

$$\psi_5{}^z = \frac{R^2}{EI_2}(\lambda-\sin\lambda)$$

$$\psi_6{}^z = \frac{R}{EI_z}\lambda$$

Equation (a) describes the relationship between displacements and corresponding forces at node i.

If the conditions,

$$q_1 = 1, q_2 = q_6 = 0; \qquad q_2 = 1, q_1 = q_6 = 0; \qquad q_6 = 1, q_1 = q_2 = 0;$$

are imposed on Eq. (a), one at a time, and corresponding forces are calculated in each case, it will yield the following stiffness coefficients,

$$\begin{aligned}\bar{S}_{11} &= (\psi_4{}^z\psi_6{}^z-\psi_5{}^z\psi_5{}^z)/\Delta^z\\ \bar{S}_{21} &= (\psi_3{}^z\psi_5{}^z-\psi_2{}^z\psi_6{}^z)/\Delta^z\\ \bar{S}_{61} &= (\psi_2{}^z\psi_5{}^z-\psi_3{}^z\psi_4{}^z)/\Delta^z\\ \bar{S}_{22} &= (\psi_1{}^z\psi_6{}^z-\psi_3{}^z\psi_3{}^z)/\Delta^z\\ \bar{S}_{62} &= (\psi_2{}^z\psi_3{}^z-\psi_1{}^z\psi_5{}^z)/\Delta^z\\ \bar{S}_{66} &= (\psi_1{}^z\psi_4{}^z-\psi_2{}^z\psi_2{}^z)/\Delta^z\end{aligned} \tag{3.40}$$

where

$$\begin{aligned}\Delta^z = \psi_1{}^z(\psi_4{}^z\psi_6{}^z-\psi_5{}^z\psi_5{}^z)-\psi_2{}^z(\psi_2{}^z\psi_6{}^z-\psi_3{}^z\psi_5{}^z)\\ +\psi_3{}^z(\psi_2{}^z\psi_5{}^z-\psi_3{}^z\psi_4{}^z)\end{aligned} \tag{3.41}$$

From equilibrium equations of statics (note that $S_{ij} = S_{ji}$), we have,

$$\begin{aligned}\bar{S}_{71} &= -\bar{S}_{11}\\ \bar{S}_{81} &= -\bar{S}_{21}\\ \bar{S}_{12,1} &= -\bar{S}_{11}R\sin\lambda-\bar{S}_{21}R(1-\cos\lambda)-\bar{S}_{61}\\ \bar{S}_{72} &= -\bar{S}_{21}\\ \bar{S}_{82} &= -\bar{S}_{22}\\ \bar{S}_{12,2} &= -\bar{S}_{21}R\sin\lambda-\bar{S}_{22}R(1-\cos\lambda)-\bar{S}_{62}\\ \bar{S}_{76} &= -\bar{S}_{61}\end{aligned} \tag{3.42}$$

$$
\begin{aligned}
\bar{S}_{86} &= -\bar{S}_{62} \\
\bar{S}_{12,6} &= -\bar{S}_{61}R \sin\lambda - \bar{S}_{62}R(1-\cos\lambda) - \bar{S}_{66} \\
\bar{S}_{77} &= -\bar{S}_{71} \\
\bar{S}_{87} &= -\bar{S}_{72} \\
\bar{S}_{12,7} &= -\bar{S}_{71}R \sin\lambda - \bar{S}_{72}R(1-\cos\lambda) - \bar{S}_{76} \\
\bar{S}_{88} &= -\bar{S}_{82} \\
\bar{S}_{12,8} &= -\bar{S}_{81}R \sin\lambda - \bar{S}_{82}R(1-\cos\lambda) - \bar{S}_{86} \\
\bar{S}_{87} &= -\bar{S}_{71} \\
\bar{S}_{12,12} &= -\bar{S}_{12,1}R \sin\lambda - \bar{S}_{12,2}R(1-\cos\lambda) - \bar{S}_{12,6} \\
\bar{S}_{12,8} &= -\bar{S}_{12,2} \\
\bar{S}_{12,7} &= -\bar{S}_{12,1}
\end{aligned}
\tag{3.42}
$$

To populate the rest of the stiffness matrix we shall now consider the displacements in the $\bar{x}\bar{z}$ plane, i.e., q_i ($i = 3, 4, 5, 9, 10,$ and 11).

For this case the expressions for S, M, V, and T in conjunction with Eq. (3.36) in terms of forces at node i can be obtained (*see* Fig. 3.19) as follows,

$$
\begin{aligned}
S(\theta) &= 0 \\
M(\theta) &= -Q_4 \cos\theta - Q_5 \sin\theta + Q_3 R \sin\theta \\
T(\theta) &= Q_4 \sin\theta - Q_5 \cos\theta - Q_3 R(1-\cos\theta) \\
V(\theta) &= -Q_3 \sin\theta
\end{aligned}
\tag{3.43}
$$

Substituting Eq. (3.43) into Eq. (3.36), integrating, and then expanding for Q_i ($i = 3, 4,$ and 5) yield,

$$
\begin{bmatrix} q_3 \\ q_4 \\ q_5 \end{bmatrix} = \begin{bmatrix} \psi_6{}^y & \psi_3{}^y & \psi_5{}^y \\ \psi_3{}^y & \psi_1{}^y & \psi_2{}^y \\ \psi_5{}^y & \psi_2{}^y & \psi_4{}^y \end{bmatrix} \begin{bmatrix} Q_3 \\ Q_4 \\ Q_5 \end{bmatrix} \tag{b}
$$

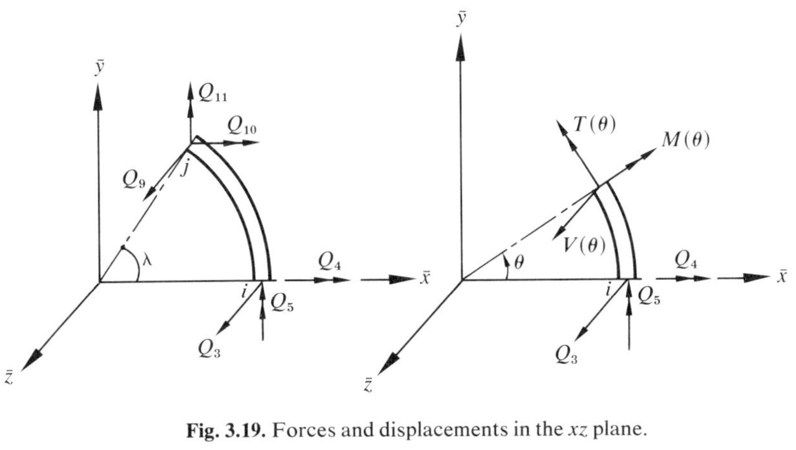

Fig. 3.19. Forces and displacements in the xz plane.

where

$$
\begin{aligned}
\psi_1{}^y &= \left(\frac{R}{GJ}+\frac{R}{EI_y}\right)\frac{\lambda}{2}+\left(\frac{R}{EI_y}-\frac{R}{GJ}\right)\frac{\sin 2\lambda}{4} \\
\psi_2{}^y &= \left(\frac{R}{EI_y}-\frac{R}{GJ}\right)\frac{\sin^2\lambda}{2} \\
\psi_3{}^y &= \left(\frac{R^2}{GJ}-\frac{R^2}{EI_y}\right)\frac{\sin^2\lambda}{2}+\frac{R^2}{GJ}(\cos\lambda-1) \\
\psi_4{}^y &= \left(\frac{R}{EI_y}+\frac{R}{GJ}\right)\frac{\lambda}{2}+\left(\frac{R}{GJ}-\frac{R}{EI_y}\right)\frac{\sin 2\lambda}{4} \\
\psi_5{}^y &= \left(\frac{R^2}{EI_y}+\frac{R^2}{GJ}\right)\frac{\lambda}{2}+\left(\frac{R^2}{EI_y}-\frac{R^2}{GJ}\right)\frac{\sin 2\lambda}{4}+\frac{R^2}{GJ}\sin\lambda \\
\psi_6{}^y &= \left(\frac{R^3}{EI_y}+\frac{R^3}{GJ}\right)\frac{\lambda}{2}+\left(\frac{R^3}{GJ}-\frac{R^3}{EI_y}\right)\frac{\sin 2\lambda}{4}+\frac{R^3}{GJ}(\lambda-2\sin\lambda)+\frac{R\lambda}{AG}
\end{aligned}
\qquad (3.44)
$$

If the conditions,

$$q=1,\, q=q=0; \qquad q=1,\, q=q=0; \qquad q=1,\, q=q=0;$$

are imposed on Eq. (b) one at a time, result forces are calculated for each condition, and the equilibrium equations of statics are utilized, will yield the following stiffness coefficient,

$$
\begin{aligned}
\bar{S}_{33} &= (\psi_1{}^y\psi_4{}^y-\psi_2{}^y\psi_2{}^y)/\Delta^y \\
\bar{S}_{43} &= (\psi_2{}^y\psi_5{}^y-\psi_3{}^y\psi_4{}^y)/\Delta^y \\
\bar{S}_{53} &= (\psi_2{}^y\psi_3{}^y-\psi_1{}^y\psi_5{}^y)/\Delta^y \\
\bar{S}_{10,3} &= -\bar{S}_{43} \\
\bar{S}_{11,3} &= -\bar{S}_{53} \\
\bar{S}_{93} &= -\bar{S}_{43}R\sin\lambda-\bar{S}_{53}R(1-\cos\lambda)-\bar{S}_{33} \\
\bar{S}_{44} &= (\psi_4{}^y\psi_6{}^y-\psi_5{}^y\psi_5{}^y)/\Delta^y \\
\bar{S}_{54} &= (\psi_3{}^y\psi_5{}^y-\psi_2{}^y\psi_6{}^y)/\Delta^y \\
\bar{S}_{94} &= -\bar{S}_{44}R\sin\lambda-\bar{S}_{54}R(1-\cos\lambda)-\bar{S}_{43} \\
\bar{S}_{10,4} &= -\bar{S}_{44} \\
\bar{S}_{11,4} &= -\bar{S}_{54} \\
\bar{S}_{55} &= (\psi_1{}^y\psi_6{}^y-\psi_3{}^y\psi_3{}^y)/\Delta^y \\
\bar{S}_{95} &= -\bar{S}_{54}R\sin\lambda-\bar{S}_{55}R(1-\cos\lambda)-\bar{S}_{53} \\
\bar{S}_{10,5} &= -\bar{S}_{54} \\
\bar{S}_{11,5} &= -\bar{S}_{55} \\
\bar{S}_{99} &= -\bar{S}_{94}R\sin\lambda-\bar{S}_{10,5}R(1-\cos\lambda)-\bar{S}_{93} \\
\bar{S}_{10,9} &= -\bar{S}_{10,4}R\sin\lambda-\bar{S}_{10,5}R(1-\cos\lambda)-\bar{S}_{10,3} \\
\bar{S}_{11,9} &= -\bar{S}_{11,4}R\sin\lambda-\bar{S}_{11,5}R(1-\cos\lambda)-\bar{S}_{11,3} \\
\bar{S}_{10,10} &= \bar{S}_{44} \\
\bar{S}_{11,10} &= \bar{S}_{54} \\
\bar{S}_{11,11} &= \bar{S}_{55}
\end{aligned}
\qquad (3.45)
$$

where

$$\Delta^y = \psi_1{}^y(\psi_4{}^y\psi_6{}^y - \psi_5{}^y\psi_5{}^y) - \psi_2{}^y(\psi_2{}^y\psi_6{}^y - \psi_3{}^y\psi_5{}^y) + \psi_3{}^y(\psi_2{}^y\psi_5{}^y - \psi_3{}^y\psi_4{}^y) \tag{3.46}$$

Hence, the three-dimensional curved beam element stiffness matrix equation in the local coordinate system can be written as follows,

$$\begin{bmatrix} F_i^{\bar{x}} \\ F_i^{\bar{y}} \\ F_i^{\bar{z}} \\ M_i^{\bar{x}} \\ M_i^{\bar{y}} \\ M_i^{\bar{z}} \\ F_j^{\bar{x}} \\ F_j^{\bar{y}} \\ F_j^{\bar{z}} \\ M_j^{\bar{x}} \\ M_j^{\bar{y}} \\ M_j^{\bar{z}} \end{bmatrix} = [\bar{S}] \begin{bmatrix} \delta_i^{\bar{x}} \\ \delta_i^{\bar{y}} \\ \delta_i^{\bar{z}} \\ \theta_i^{\bar{x}} \\ \theta_i^{\bar{y}} \\ \theta_i^{\bar{z}} \\ \delta_j^{\bar{x}} \\ \delta_j^{\bar{y}} \\ \delta_j^{\bar{z}} \\ \theta_j^{\bar{x}} \\ \theta_j^{\bar{y}} \\ \theta_j^{\bar{z}} \end{bmatrix} \tag{3.47}$$

where the stiffness coefficients, S_{ij}, are defined in Eqs. (3.40–3.45).

In case of the two-dimensional curved beam element, Eq. (3.47) reduces to,

$$\begin{bmatrix} F_i^{\bar{x}} \\ F_i^{\bar{y}} \\ M_i^{\bar{z}} \\ F_j^{\bar{x}} \\ F_j^{\bar{y}} \\ M_j^{\bar{z}} \end{bmatrix} = \begin{bmatrix} \bar{S}_{11} & & & \text{Sym.} & & \\ \bar{S}_{21} & \bar{S}_{22} & & & & \\ \bar{S}_{61} & \bar{S}_{62} & \bar{S}_{66} & & & \\ \bar{S}_{71} & \bar{S}_{72} & \bar{S}_{76} & \bar{S}_{77} & & \\ \bar{S}_{81} & \bar{S}_{82} & \bar{S}_{86} & \bar{S}_{87} & \bar{S}_{88} & \\ \bar{S}_{12,1} & \bar{S}_{12,2} & \bar{S}_{12,6} & \bar{S}_{12,7} & \bar{S}_{12,8} & \bar{S}_{12,12} \end{bmatrix} \begin{bmatrix} \delta_i^{\bar{x}} \\ \delta_i^{\bar{y}} \\ \theta_i^{\bar{x}} \\ \delta_j^{\bar{x}} \\ \delta_j^{\bar{y}} \\ \theta_j^{\bar{z}} \end{bmatrix} \tag{3.48}$$

where the stiffness coefficients are defined in Eqs. (3.40) and (3.45).

The stiffness matrices in Eqs. (3.47) and (3.48) are limited in their application to curved beam elements that are oriented as shown in Fig. 3.17. To accommodate for elements which are arbitrarily oriented, a transformation matrix is developed. This will allow for three rigid body rotations of the beam element:

1. Rotation about the y-axis in the xz-plane.
2. Rotation about the z_1-axis in the x_1y_1 plane.
3. Rotation about the x_2-axis in the y_2z_2 plane.

Figure 3.20 illustrates such rigid body rotations.

The rotations shown in Fig. 3.20(a) and (b) are identical to those previously derived (*see* Eq. 3.10) and hence, the xyz and $x_2y_2z_2$ coordinates are related as follows,

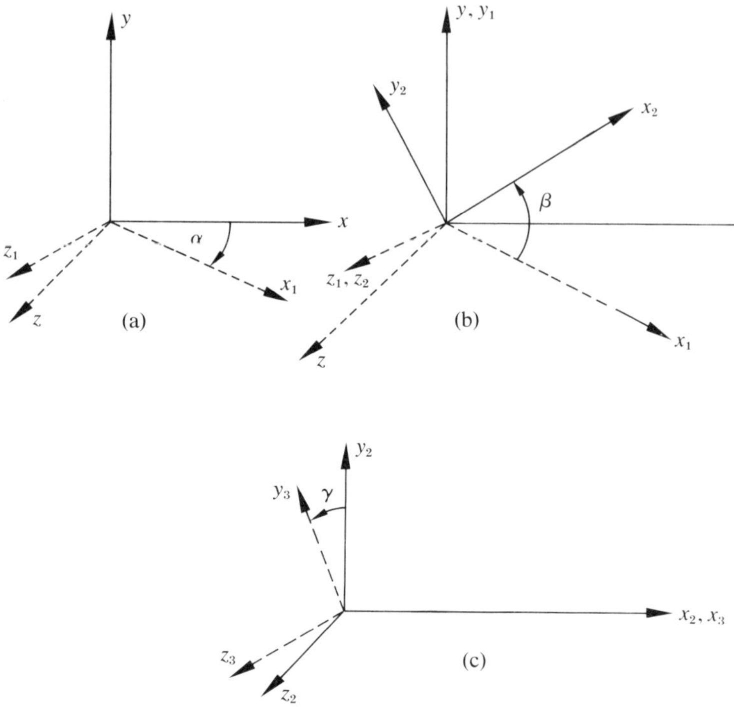

Fig. 3.20. Coordinate transformations.

$$\begin{bmatrix} x_2 \\ y_2 \\ z_2 \end{bmatrix} = \begin{bmatrix} \cos\alpha\cos\beta & \sin\beta & \cos\beta\sin\alpha \\ -\cos\alpha\sin\beta & \cos\beta & -\sin\alpha\sin\beta \\ -\sin\alpha & 0 & \cos\alpha \end{bmatrix} \begin{bmatrix} x \\ y \\ z \end{bmatrix} \tag{3.9}$$

In a similar manner, the relationship between the $x_2y_2z_2$ and $x_3y_3z_3$ coordinates for a rotation, γ, about the x_2-axis can be expressed as follows,

$$\begin{bmatrix} x_3 \\ y_3 \\ z_3 \end{bmatrix} = \begin{bmatrix} 1 & 0 & 0 \\ 0 & \cos\gamma & \sin\gamma \\ 0 & -\sin\gamma & \cos\gamma \end{bmatrix} \begin{bmatrix} x_2 \\ y_2 \\ \gamma_2 \end{bmatrix} \tag{3.49}$$

Upon substituting Eq. (3.10) into Eq. (3.49) and carrying out matrix multiplication, the following is obtained,

$$\begin{bmatrix} x_3 \\ y_3 \\ z_3 \end{bmatrix} = [T] \begin{bmatrix} x \\ y \\ z \end{bmatrix} \tag{3.50}$$

where $[T]$ is given by,

$$[T]=\begin{bmatrix} (\cos\alpha\cos\beta) & (\sin\beta) & (\cos\beta\sin\alpha) \\ -(\sin\alpha\sin\gamma+ & (\cos\beta\cos\gamma) & (\sin\gamma\cos\alpha- \\ \quad\cos\alpha\sin\beta\cos\gamma) & & \quad\sin\alpha\sin\beta\cos\gamma) \\ -(\sin\alpha\cos\gamma- & (-\sin\gamma\cos\beta) & (\cos\alpha\cos\gamma+ \\ \quad\cos\alpha\sin\beta\sin\gamma) & & \quad\sin\alpha\sin\beta\sin\gamma) \end{bmatrix} \tag{3.51}$$

Thus, in the reference coordinate system Eqs. (3.47) and (3.48) become

$$\begin{bmatrix} F_i^x \\ F_i^y \\ F_i^z \\ M_i^x \\ M_i^y \\ M_i^z \\ F_j^x \\ F_j^y \\ F_j^z \\ M_j^x \\ M_j^y \\ M_j^z \end{bmatrix} = [S]_1 \begin{bmatrix} \delta_i^x \\ \delta_i^y \\ \delta_i^z \\ \theta_i^x \\ \theta_i^y \\ \theta_i^z \\ \delta_j^x \\ \delta_j^y \\ \delta_j^z \\ \theta_j^x \\ \theta_j^y \\ \theta_j^z \end{bmatrix} \tag{3.52}$$

and

$$\begin{bmatrix} F_i^x \\ F_i^y \\ M_i^z \\ F_j^x \\ F_j^y \\ M_j^z \end{bmatrix} = [S]_2 \begin{bmatrix} \delta_i^x \\ \delta_i^y \\ \theta_i^z \\ \delta_j^x \\ \delta_j^y \\ \theta_j^z \end{bmatrix} \tag{3.53}$$

respectively.

The stiffness matrices in Eqs. (3.52) and (3.53) are defined respectively as follows,

$$[S]_1=\begin{bmatrix} [T] & 0 & 0 & 0 \\ 0 & [T] & 0 & 0 \\ 0 & 0 & [T] & 0 \\ 0 & 0 & 0 & [T] \end{bmatrix}^{-1} [\bar{S}]_1 \begin{bmatrix} [T] & 0 & 0 & 0 \\ 0 & [T] & 0 & 0 \\ 0 & 0 & [T] & 0 \\ 0 & 0 & 0 & [T] \end{bmatrix} \tag{3.54}$$

and

$$[S]_2=\begin{bmatrix} [T] & 0 \\ 0 & [T] \end{bmatrix}^{-1} [\bar{S}]_2 \begin{bmatrix} [T] & 0 \\ 0 & [T] \end{bmatrix} \tag{3.55}$$

3.12 STRUCTURES WITH DISTRIBUTED LOADS

The basis of the stiffness matrix method is that it relates discrete nodal forces to corresponding discrete nodal displacements. Hence, any load, concentrated or distributed, which is not applied or lumped at the chosen nodal points of the structure cannot be implemented by the method. To render its application to such structures, the non-nodal loads will be converted into nodal equivalent loads by utilizing the principles of superposition.

Figure 3.21(a) shows a one story building frame structure acted upon by a distributed load and a concentrated load which are not applied at the chosen node points. Utilizing the principle of superposition the forces and displacements of this structure are equivalent to the sum of the forces and displacements of the same structure loaded as shown in Fig. 3.21 (b) and (c).

In Fig. 3.21(b) the displacements at all node points are fully restrained. The restraining fixed end forces can be easily calculated by simple beam formulas and are considered as temporarily external applied forces. The fixed end forces do not actually exist and hence should be removed by applying forces of equal magnitude but in the opposite direction at the respective node points as shown in Fig. 3.21(c). Once this is done, the structure in Fig. 3.21(c) can be analyzed by the stiffness matrix method. The final solution for the structure in Fig. 3.21(a) can then be obtained by superimposing displacements and internal forces obtained from the analyses of Fig. 3.21(b) and (c).

The above procedure applies to any structure having any distributed load as illustrated in Fig. 3.22.

Notice that the fixed support at node 1 can absorb both bending

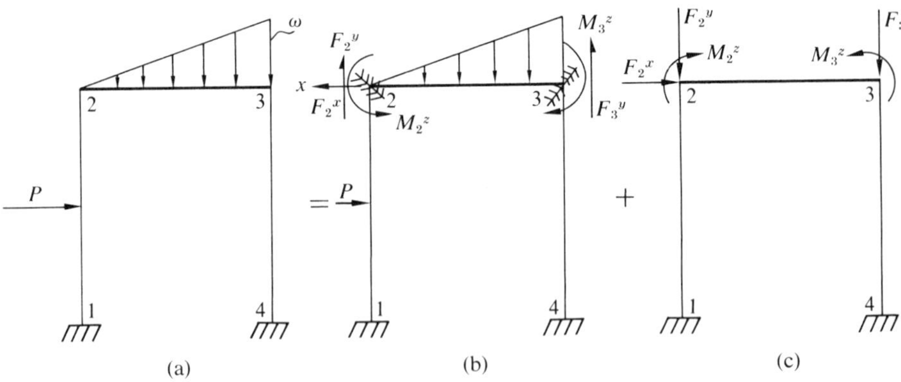

Fig. 3.21. Structures with loads not applied at the nodes.

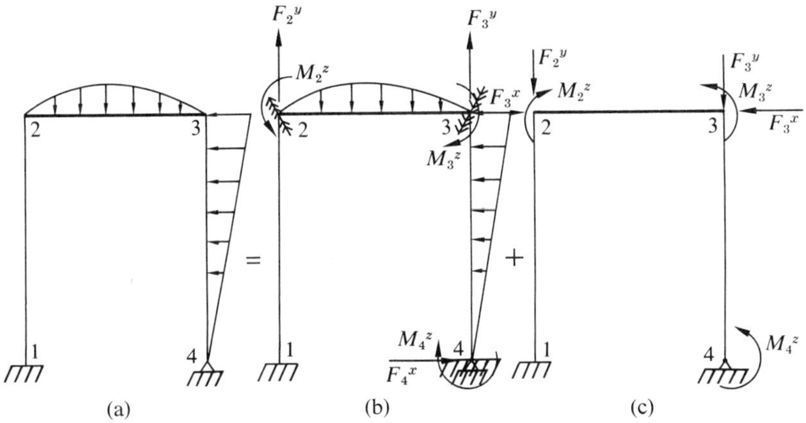

Fig. 3.22. Structure with distributed load.

moments and forces, which indicates that "joint relaxation" at that point is not required. Therefore, the fixed end forces at node 1 need not be shown in Fig. 3.22(c). On the other hand, node 4 is simply supported, which can absorb forces but not bending moments, and therefore a fixed end bending moment at that point must be introduced as shown in Fig. 3.22(c).

Quite often, beam elements with distributed loads do not necessarily lie in the reference coordinate system, for instance, as shown in Fig. 3.23(a). In this case, once the fixed end forces are calculated in the local

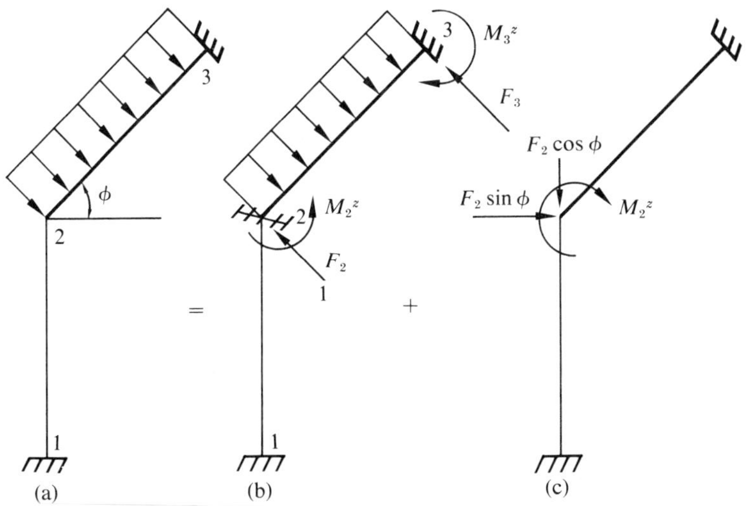

Fig. 3.23. Frame structure with distributed load.

coordinate system of the beam element under consideration, they should be transformed into the reference coordinate system of the structure. (*See* Fig. 3.23(c) for illustration.)

In general (*see* Eq. 3.10), the beam element forces in the local coordinate system are related to those in the reference coordinates as follows,

$$\{\bar{Q}\} = [T]\{Q\}, \tag{3.55a}$$

or

$$\{Q\} = [T]^{-1}\{\bar{Q}\}. \tag{3.56}$$

For the beam element Eq. (3.56) can be written as,

$$\begin{bmatrix} F_i^x \\ F_i^y \\ F_i^z \\ M_i^x \\ M_i^y \\ M_i^z \\ F_j^z \\ F_j^y \\ F_j^z \\ M_j^x \\ M_j^y \\ M_j^z \end{bmatrix} = \begin{bmatrix} [T] & & & \\ & [T] & & \\ & & [T] & \\ & & & [T] \end{bmatrix}^{-1} \begin{bmatrix} F_i^{\bar{x}} \\ F_i^{\bar{y}} \\ F_i^{\bar{z}} \\ M_i^{\bar{x}} \\ M_i^{\bar{y}} \\ M_i^{\bar{z}} \\ F_j^{\bar{x}} \\ F_j^{\bar{y}} \\ F_j^{\bar{z}} \\ M_j^{\bar{x}} \\ M_j^{\bar{y}} \\ M_j^{\bar{z}} \end{bmatrix} \tag{3.57}$$

in which $[T]$ is defined in Eq. (3.17a).

3.13 STRUCTURES WITH THERMAL LOADS

The treatment of thermally loaded structures is analogous to that of structures with distributed loads. Thus, all structural elements which are thermally loaded have their nodes initially locked against displacement and the fixed end forces resulting from such loads are calculated. Subsequently, these forces are then applied at their respective node points with reversal of sign. Thus if we let $\{Q\}$ represent the applied external forces and $\{Q_T\}$ the forces resulting from thermal effects, then Eq. (3.1a) can be rewritten as,

$$\{Q\} - \{Q_T\} = [S]\{q\} \tag{3.58}$$

or

$$\{Q\} = [S]\{q\} + \{Q_T\}. \tag{3.58a}$$

For an axial rod element in local coordinates. Eq. (3.58a) can be written in an expanded form as,

$$\begin{bmatrix} F_i^{\bar{x}} \\ F_j^{\bar{x}} \end{bmatrix} = \frac{AE}{L}\begin{bmatrix} 1 & -1 \\ 1 & 1 \end{bmatrix}\begin{bmatrix} \delta_i^{\bar{x}} \\ \delta_j^{\bar{x}} \end{bmatrix} + AE\mu T\begin{bmatrix} 1 \\ -1 \end{bmatrix} \tag{3.59}$$

where μ = coefficient of thermal expansion,
T = change in temperature above a given datum temperature,
A = rod cross-sectional area,
L = rod length,
E = Young's Modulus.

For a beam element, Eq. (3.58a) may be written as,

$$\begin{bmatrix} F_i^{\bar{x}} \\ F_i^{\bar{y}} \\ F_i^{\bar{z}} \\ M_i^{\bar{x}} \\ M_i^{\bar{y}} \\ M_i^{\bar{z}} \\ F_j^{\bar{x}} \\ F_j^{\bar{y}} \\ F_j^{\bar{z}} \\ M_j^{\bar{x}} \\ M_j^{\bar{y}} \\ M_j^{\bar{z}} \end{bmatrix} = [S]\{q\} + \begin{bmatrix} F_{iT}^{\bar{x}} \\ 0 \\ 0 \\ 0 \\ M_{iT}^{\bar{y}} \\ -M_{iT}^{\bar{z}} \\ -F_{jT}^{\bar{x}} \\ 0 \\ 0 \\ 0 \\ -M_{jT}^{\bar{y}} \\ M_{jT}^{\bar{z}} \end{bmatrix} \tag{3.60}$$

in which

$$F_T^{\bar{x}} = \mu\int_A ETdA, \quad M_T^{\bar{y}} = \mu\int_A ETzdA. \quad M_T^{\bar{z}} = \mu\int_A ETydA. \tag{3.61}$$

where μ = coefficient of thermal expansion,
E = Young's Modulus,
T = temperature change.

Illustrative Examples

Example 10

For the beam structure shown find all node point rotations by the stiffness matrix method.

$$E = 30\times 10^6, \qquad A = 2\text{ in}^2, \qquad I = 0.67\text{ in}^4.$$

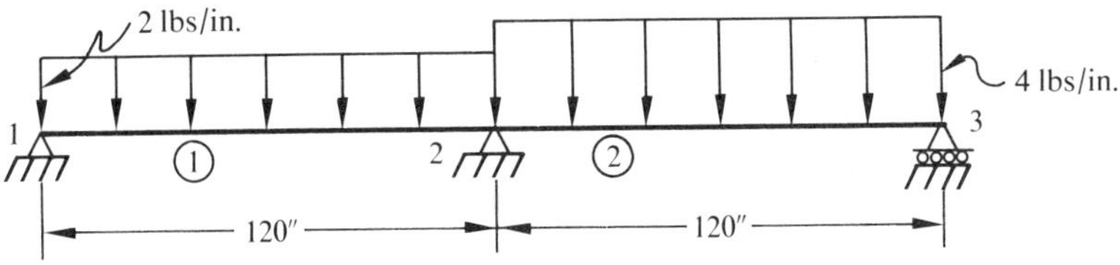

To obtain a solution for the above example the distributed load has to be resolved into loads applied at the nodes as was discussed in Section 3.12. Thus, the given structure can be replaced by the following,

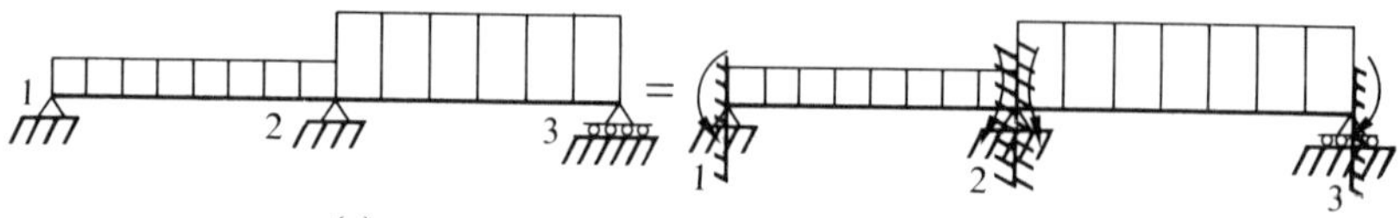

(a) (b)

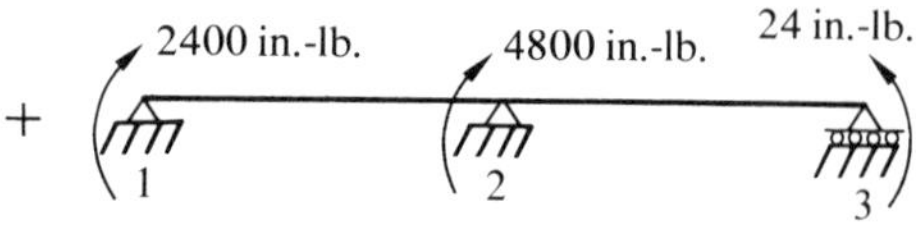

(c)

Applying the stiffness matrix method to the structure in (c) yields the following aggregate stiffness matrix equation,

$$\begin{bmatrix} -2400 \\ 0 \\ -2400 \\ 0 \\ 4800 \end{bmatrix} = 10^5 \begin{bmatrix} 6.7 & 0 & 3.35 & 0 & 0 \\ 0 & 10 & 0 & -5 & 0 \\ 3.35 & 0 & 13.4 & 0 & 3.35 \\ 0 & -5 & 0 & 5 & 0 \\ 0 & 0 & 3.35 & 0 & 6.7 \end{bmatrix} \begin{bmatrix} \theta_1^z \\ \delta_2^x \\ \theta_2^z \\ \delta_3^x \\ \theta_3^z \end{bmatrix}$$

Solving for the displacements yields,

$$\begin{bmatrix} \theta_1^z \\ \delta_2^x \\ \theta_2^z \\ \delta_3^x \\ \theta_3^z \end{bmatrix} = 10^{-7} \begin{bmatrix} 17.4 & 0 & -5 & 0 & 2.5 \\ 0 & 20 & 0 & 20 & 0 \\ -5 & 0 & 10 & 0 & -5 \\ 0 & 20 & 0 & 40 & 0 \\ 25 & 0 & -5 & 0 & 17 \end{bmatrix} \begin{bmatrix} -2400 \\ 0 \\ -2400 \\ 0 \\ 4800 \end{bmatrix}$$

or

$$\theta_1^z = -1.79 \times 10^{-3} \text{ rad}$$
$$\delta_2^x = 0$$
$$\theta_2^z = -3.58 \times 10^{-3} \text{ rad}$$
$$\delta_3^x = 0$$

and

$$\theta_3^z = 8.96 \times 10^{-3} \text{ rad}$$

Example 11

Find the internal loads and nodal displacements of the plane truss structure shown.

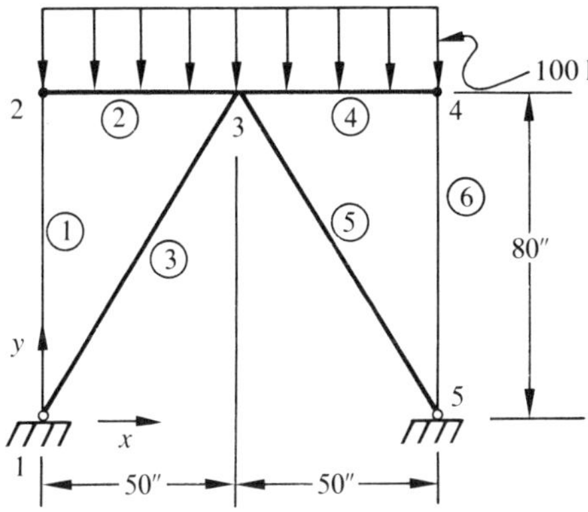

Before we can apply the matrix technique in this particular example the distributed load must be converted into equivalent concentrated forces located at specified node points. This is done by fixing joints 2, 3, and 4, calculating the resulting end restrained forces for members 2 and 4, then applying these forces with reverse in sign at their respective nodes. Thus, the node forces become, $F_2^y = -2500$ lbs. $F_3^y = -5000$ lbs and $F_4^y = -2500$ lbs. Notice that bending moments are not considered due to the fact that pinned joints cannot resist such loads.

Taking advantage of symmetry, both in loading and construction of the truss, about a vertical line through node 3, only half of the structure need to be considered. Notice that node 3 can translate vertically only. Hence by considering the left half of the structure node 3 would act as if being on a roller support.

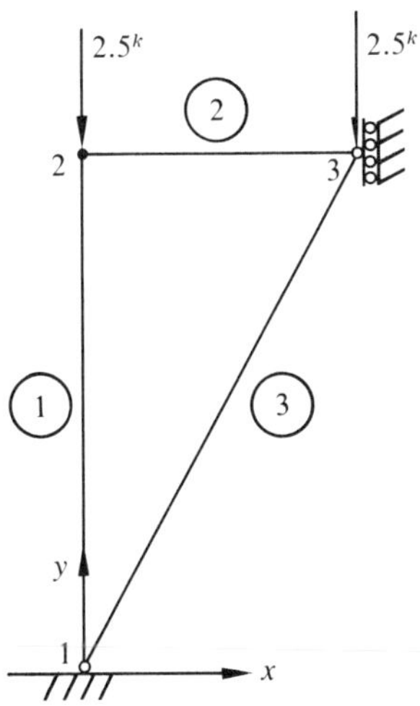

Node Coordinates			Rod Length	
Node	x	y	Rod	Length
1	0	0	1	80
2	0	80	2	50
3	50	80	3	94.3

Coefficients of Transformation Matrix.

Rod	Coefficients	
1	$T_{11} = 0,$	$T_{12} = 1$
2	$T_{11} = 1,$	$T_{12} = 0$
3	$T_{11} = 0.53,$	$T_{12} = 0.85$

The individual element stiffness matrices are,

Rod 1

$$[S]_1 = \frac{AE}{94.3}\begin{bmatrix} 0 & & \text{Sym.} & \\ 0 & 1.18 & & \\ 0 & 0 & 0 & \\ 0 & -1.18 & 0 & 1.18 \end{bmatrix}$$

Rod 2

$$[S]_2 = \frac{AE}{94.3}\begin{bmatrix} 1.88 & & \text{Sym.} & \\ 0 & 0 & & \\ -1.88 & 0 & 1.88 & \\ 0 & 0 & 0 & 0 \end{bmatrix}$$

Rod 3

$$[S]_3 = \frac{AE}{94.3}\begin{bmatrix} 0.24 & & \text{Sym.} & \\ 0.45 & 0.72 & & \\ -0.28 & -0.45 & 0.28 & \\ -0.45 & -0.72 & 0.45 & 0.72 \end{bmatrix}$$

The aggregate stiffness matrix equation for the structure is,

$$\begin{bmatrix} F_1^x \\ F_1^y \\ F_2^x \\ F_2^y \\ F_3^x \\ F_3^y \end{bmatrix} = \frac{AE}{94.3}\begin{bmatrix} 0.28 & & & & & \\ 0.45 & 1.9 & & \text{Sym.} & & \\ 0 & 0 & 1.88 & & & \\ 0 & -1.18 & 0 & 1.18 & & \\ -0.28 & -0.45 & -1.88 & 0 & 2.16 & \\ -0.45 & -0.72 & 0 & 0 & 0.45 & 0.72 \end{bmatrix}\begin{bmatrix} \delta_1^x \\ \delta_1^y \\ \delta_2^x \\ \delta_2^y \\ \delta_3^x \\ \delta_3^y \end{bmatrix}$$

Boundary Conditions	Applied Forces	
$\delta_1^x = \delta_1^y = \delta_3^x = 0$	$F_2^y = -2500$ lbs	$F_3^y = -2500$ lbs

Deleting rows and columns corresponding to the zero displacements, the following reduced aggregate stiffness matrix equation is obtained.

$$\begin{bmatrix} -2.5 \\ 0 \\ -2.5 \end{bmatrix} = \frac{AE}{94.3}\begin{bmatrix} 1.88 & 0 & 0 \\ 0 & 1.18 & 0 \\ 0 & 0 & 0.72 \end{bmatrix}\begin{bmatrix} \delta_2^x \\ \delta_2^y \\ \delta_3^y \end{bmatrix}$$

STEVE CARR

ERRATA
Matrix Structural Analysis

p. 2: Eq. (1.5): change {A} to [A]

p. 4: Eq. (1.12a): row 2, column 2, change 4 to 5

p. 11: change $C_{21} = M_{21}$ to $C_{21} = -M_{21}$

p. 15: Line 2 from top: change L_{2m} to L^2_{2m}

p. 15: Line 4 from top: expression for L_{33} must be under the square root and change L_{3m} to L^2_{3m}

p. 18: Eq. (1.43): Left side of eq. must have superscript -1 to indicate matrix inverse

p. 27: Fig. (2.3c): change δ^z_i to θ^z_i

p. 37: Second line from top, fourth term: change xdx to x^2dx

p. 39: Eq. (2.38a): change $\frac{1}{2}{}_{jj}\, Q^2_j$ to $\frac{1}{2}\, \alpha_{jj}\, Q^2_j$

p. 42: Prob. (2.6): change $(1 - R^2_i)$ and $(1 + R^2_i)$ to $(r^2 - R_i{}^2)$ and $(r^2 + R_i{}^2)$

p. 43: Prob. (2.10): change horizontal distances on sketch to 2L each

p. 48. Eq. (3.1): change J_{j3} to S_{j3}

p. 53: Fig. (3.2d&e): change "Unit translation" to Unit rotation

p. 54: Eq. (c): move the first vertical dotted line one column to the right

p. 55: Fig. (3.4): change u_z to θ_z

p. 63: Eq. (3.15a): row 2, column 1, term should be divided by $L\zeta$

p. 66: Eq. (u): delete middle expression for T_{12}, it should be $T_{12} = -1$

p. 67: Eq. (e): add sixth row (o o 0.353 -0.353 -0.353 0.353)

p. 76: Eq. (a): interchange elements S_{21} and S_{31} with elements S_{22} and S_{32} respectively, and make the appropriate changes accordingly in the rest of the eqs.

p. 77: Eq. (3.20): row 4, column 1, change $-T^2_{12}$ to $-T_{11}^2$

p. 80: Third line from top: change xy to xz

p. 81: Eq. (i): interchange element in row 4, column 2, with that in column row 4, column 3

p. 83: Eq. (3.37): change α_6 to α^z_6

p. 86: Eq. (3.28): change α_3 to α^z_3

pgs. 87, 88, 89: Eq. (3.30a): change all T_{ij} to T_{ji}, i.e. T_{12} to T_{21}, T_{15} to T_{51} etc.

pgs. 87, 88: Eq. (3.30a): in the expressions for the following coefficients: S_{15}, S_{16}, S_{24}, S_{25}, S_{26}, S_{34} and S_{35}, interchange the (-) signs and the (+) signs and also α^y_2 and α^z_2

pgs. 88, 89: Eq. (3.30a): change $S_{48} = -S_{15}$ to $S_{48} = S_{24}$, $S_{49} = S_{16}$ to $S_{49} = S_{34}$, $S_{57} = S_{24}$ to $S_{57} = -S_{15}$, $S_{59} = -26$ to $S_{59} = -S_{35}$, $S_{67} = S_{34}$ to $S_{67} = -S_{16}$, $S_{68} = -S_{35}$ to $S_{68} = -S_{26}$

p. 89: Eq. (3.30a): Delete the expressions for S_{ij} (i = 7, 8, 9; j = 10, 11, 12) and make them read as follows:

$$S_{ij}\ (i = 7, 8, 9;\ j = 10, 11, 12) = -S_{mn}\ (m = 1, 2, 3;\ n = 4, 5, 6)$$

p. 91: Eq. (3.34a): row 4, column 1, change $T_{21}{}^2\ \alpha^z_4$ to $-T^2_{21}\ \alpha^z_4$; row 2, column 2, change $T_{22}\,\alpha^z_4$ to $T_{22}{}^2\ \alpha^z_4$; row 6, columns 2 and 4, change α_2 to $\alpha_2{}^z$; row 5, column 3, change α^x_2 to α^z_2

p. 98: Eq. (3.35): change $[+(\theta)]^2$ to $[V(\theta)]^2$

p. 99: Eq. (a): change Q_3 to Q_6

p. 125: Fig. (4.2): Equivalent beam should be:

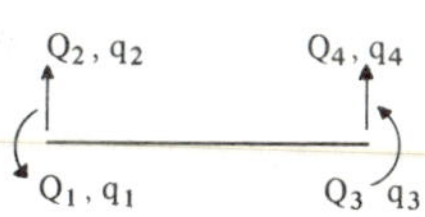

pgs. 138, 139: Eq. (4.13c): ignore bars on all S_{ij}

Solving the above equation for the displacements and then utilizing the individual element stiffness matrix equations will yield the internal forces on each rod.

3.14 SUPPLEMENTARY PROBLEMS

Problem 3.1. Given an axial rod element whose axial rigidity is given by $EA = E_0A_0e^{\beta x}$, where E_0, A_0, and β are constants, derive the element stiffness matrix equation which relates the nodal forces to the corresponding nodal displacements.

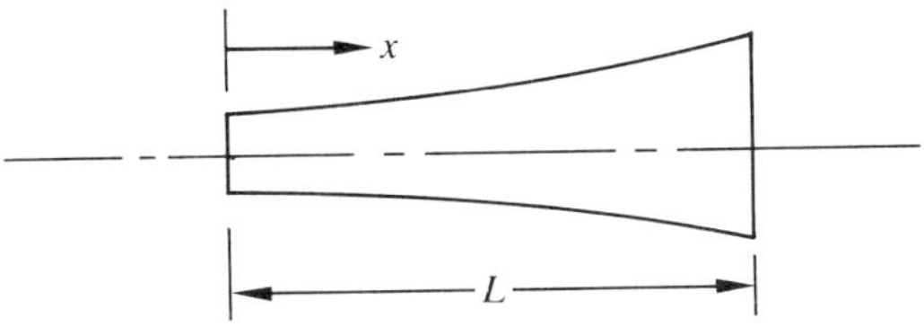

Problem 3.2. Find the stiffness matrix of the torsional bar shown, using the following differential equation,

$$M^x = -GJ\frac{d\theta^x}{dx}.$$

(a) $J = J_0 = \text{constant}$

(b) $J = J_0e^{-\alpha x}$

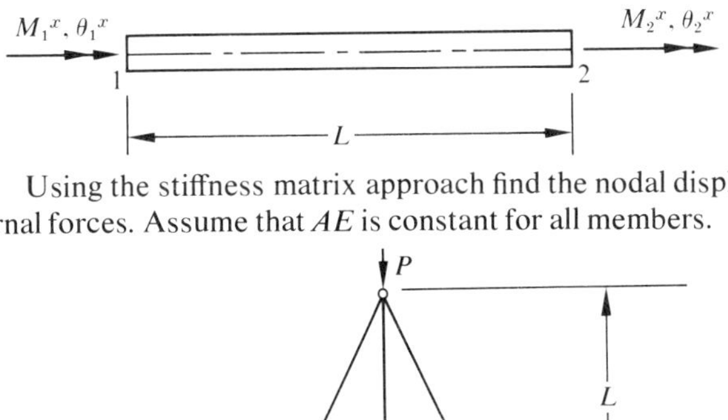

Problem 3.3. Using the stiffness matrix approach find the nodal displacements and all the internal forces. Assume that AE is constant for all members.

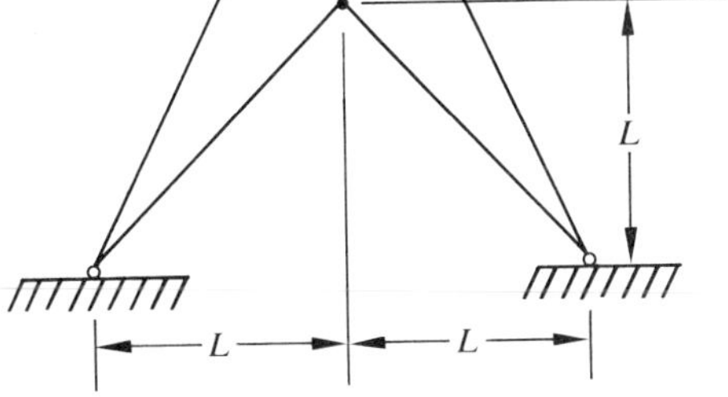

Problem 3.4. Set up the reduced aggregate stiffness matrix. Assume that AE is constant for all members.

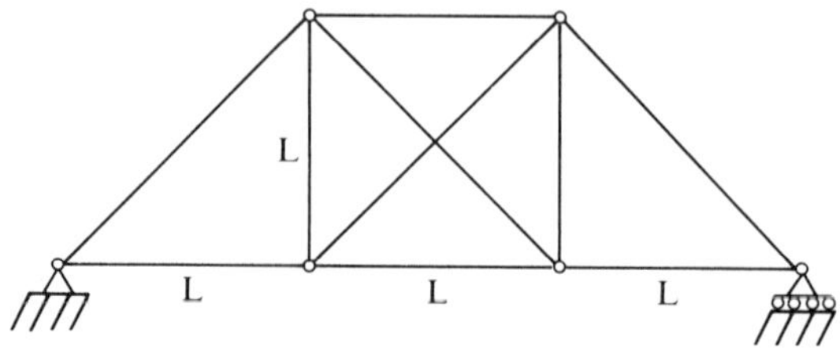

Problem 3.5. Find the internal forces on all members of the space truss-structure. $AE =$ constant for all members.

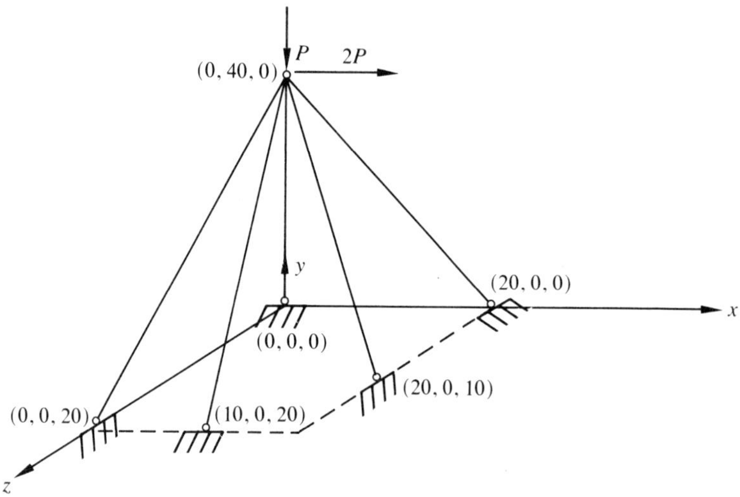

Problem 3.6. Set up the reduced aggregate stiffness matrix for the space truss shown.

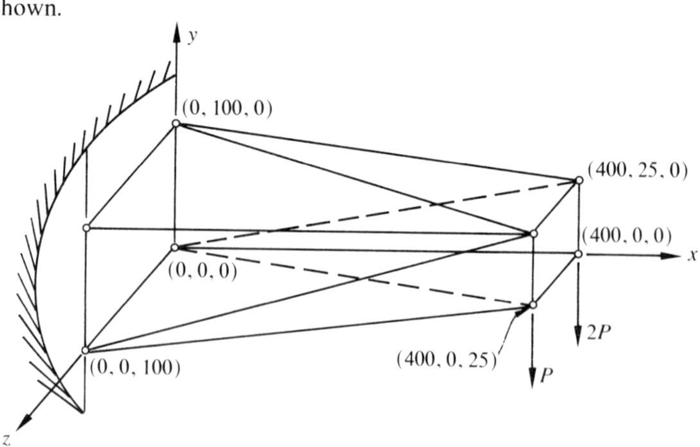

Problem 3.7. Find the nodal displacements and all member forces for the truss-structure shown. Assume AE to be constant for all members.

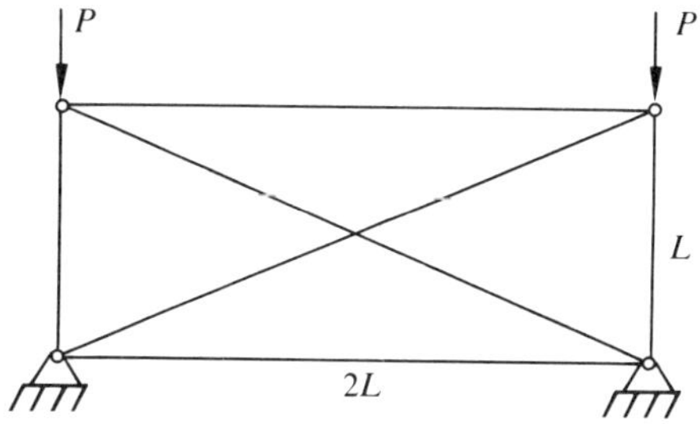

Problem 3.8. For the two-dimensional beam shown, find all unknown reactions. Neglect shear deformation.

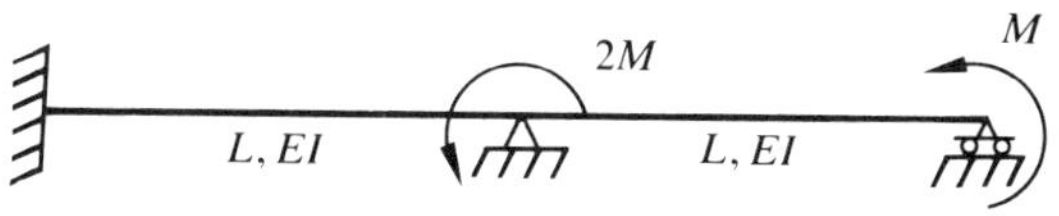

Problem 3.9. Find the displacement at the cantilevered end of the beam structure shown.

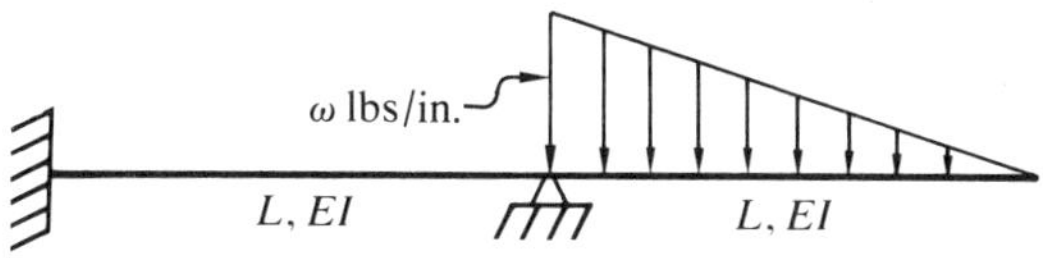

Problem 3.10. Find the tip displacements of the frame structure given below. Neglect shear deformation.

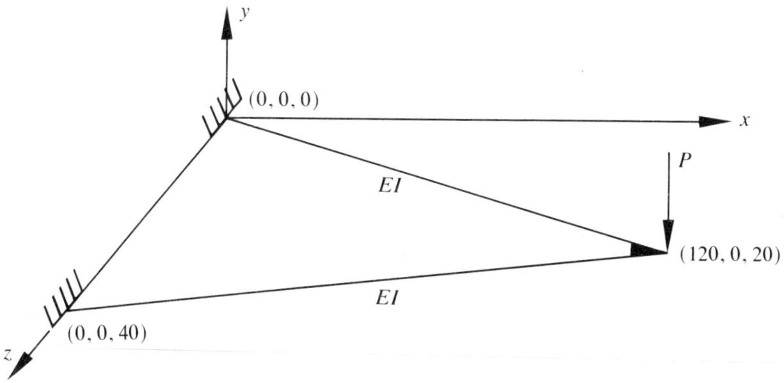

Problem 3.11. Find all internal forces on the rigid frame structure. Neglect shear deformation. $EI = 90 \times 10^6$ psi, $A = 2$ in.^{2}.

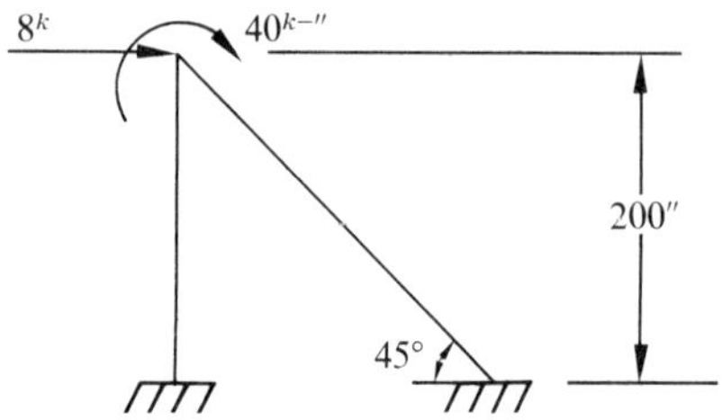

Problem 3.12. Obtain the reduced aggregate stiffness matrix for the two story building frame shown. Assume shear deformation is negligible.

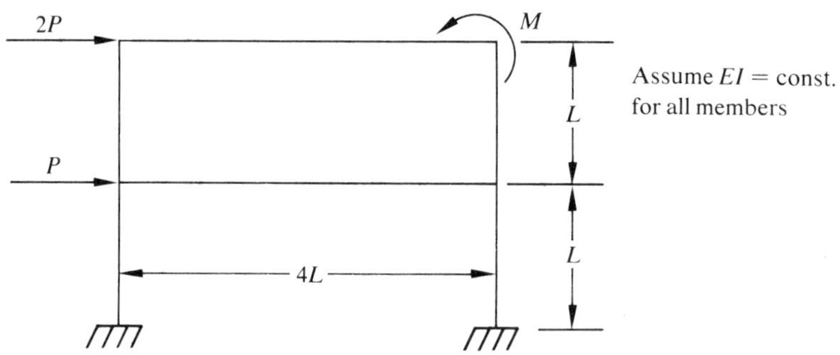

Problem 3.13. Derive the functional relationship of the form, $Q = [S]\{q\}$ for the beam structure shown. Do not use the already derived element stiffness matrices.

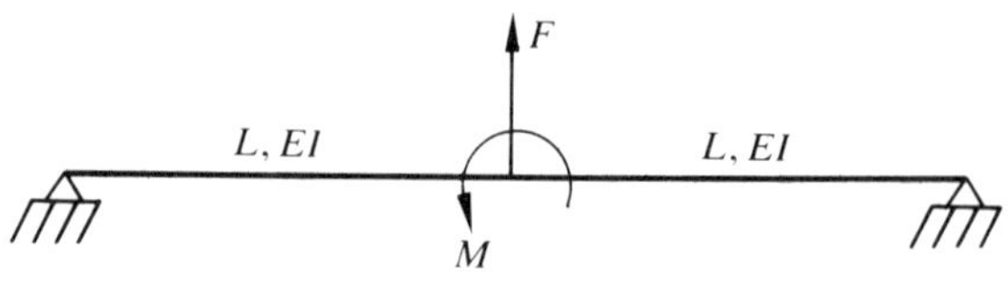

Problem 3.14. Given an L-shaped beam structure as shown, find its aggregate stiffness matrix.

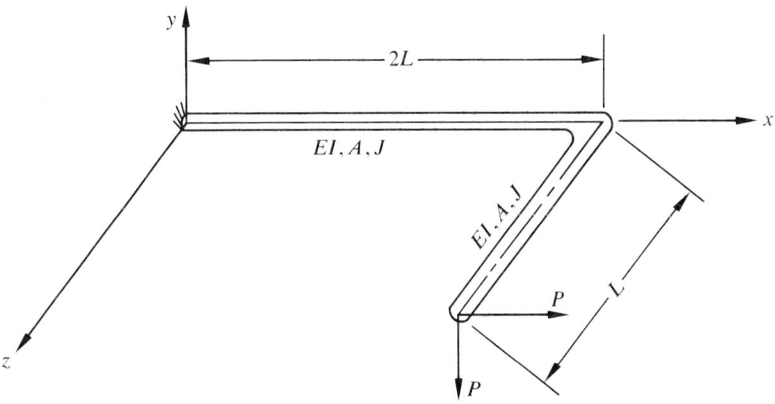

Problem 3.15. Given the circular arch as shown, find the displacements at the point of loading.

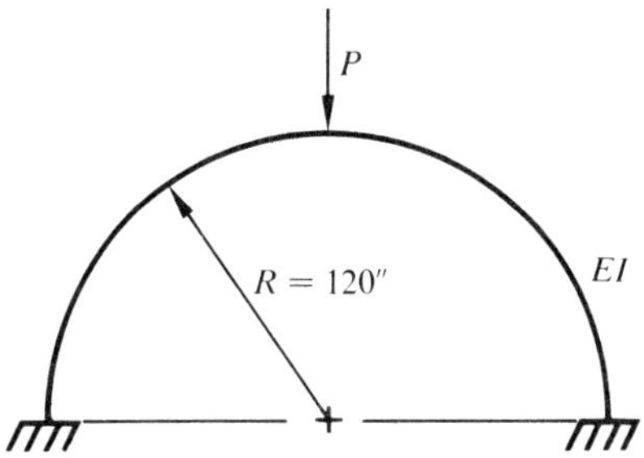

Problem 3.16. Solve Problem 3.15 by approximating the arch by; (a) two straight segments, (b) four straight segments; and compare results.

Problem 3.17. Set up the aggregate stiffness matrix for the structure shown.

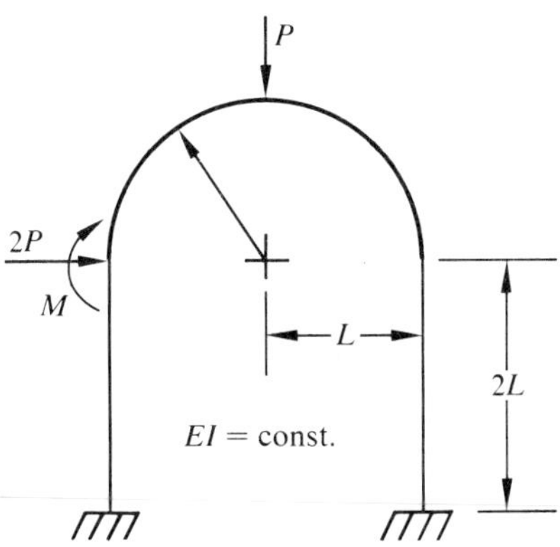

Problem 3.18. Solve for the displacements and internal forces for each of the structures shown.

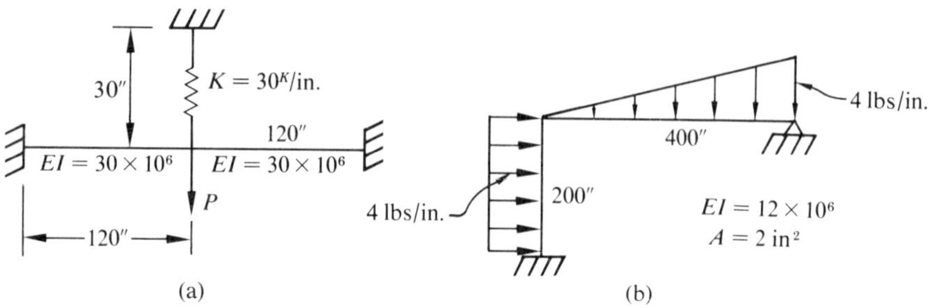

Problem 3.19. Generate the aggregate stiffness matrix of the frame structure shown without resorting to previous results. Consider bending deformations only.

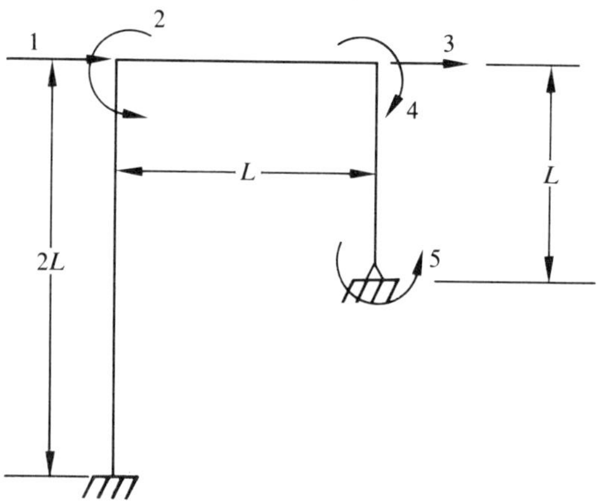

Problem 3.20. Derive directly the reduced aggregate stiffness matrix for the truss-structure shown below. Do not resort to the element stiffness matrices already derived in your text. Show all steps used in the derivation.

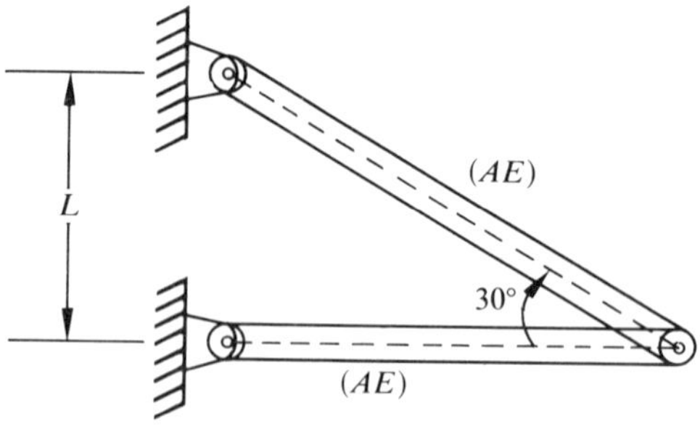

Problem 3.21. Find the displacements of the structure shown below. Neglect the effects of axial and shear deformations. Also, assume that the vertical deflection of node 2 is negligible.

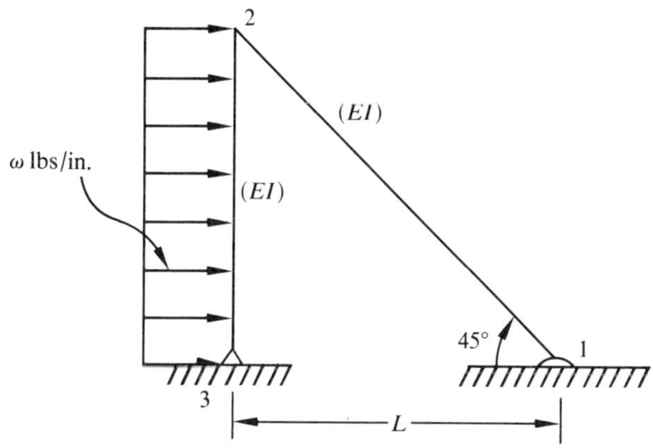

Problem 3.22. The differential equation governing the axial deformation of an axial rod is given by,

$$Q^{(x)} = -EA\frac{dq^{(x)}}{dx},$$

where EA = constant.

Derive the rod element stiffness matrix by merely utilizing the above given differential equation.

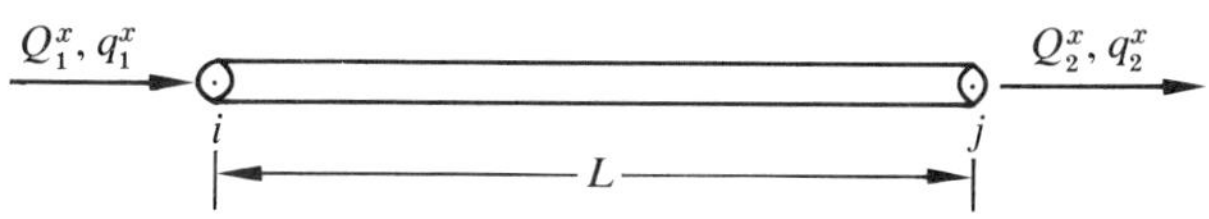

Problem 3.23. Given an axial rod element with its cross-sectional area varying linearly from A_1 to A_2, derive the stiffness matrix of the element.

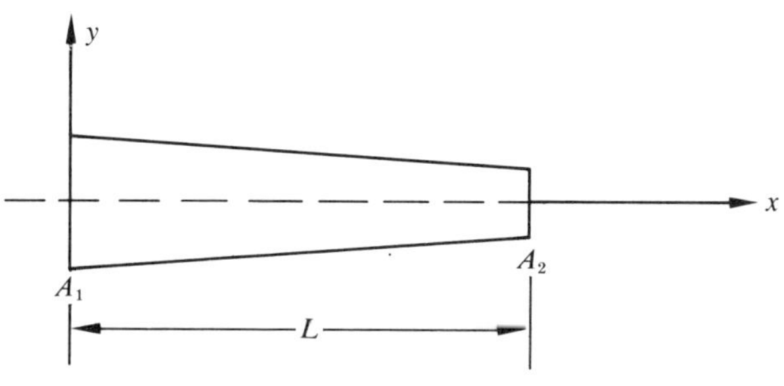

Problem 3.24. Derive the functional relationship in terms of stiffness coefficients between the coordinate forces and displacements shown in the figure below. Consider bending stresses only.
Note: Detailed Derivation Is Required.

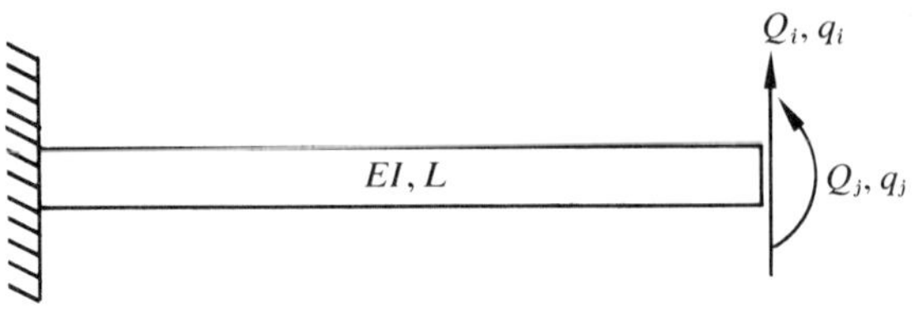

Problem 3.25. Find all nodal displacements and internal forces on beam element (1–2). Neglect shear deformation in beam and assume that its horizontal elongation is negligible.

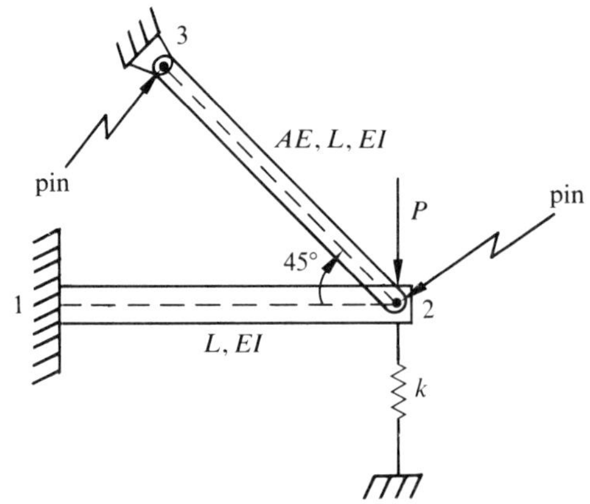

where $AE = 10^7$

$EI = 10^6$

$L = 100$ in.

$K = 10^4$ lb/in.

$P = 5000$ lbs

Problem 3.26. Solve Problem 3.25 if in addition to the load, P, member (2–3) was given a temperature rise of T°F above room temperature. α = thermal expansion coefficient of the rod.

Problem 3.27. Solve Problem 3.11, if in addition the vertical member is subjected to a temperature distribution given by,

$$T = 600\frac{y}{L}.$$

where L = vertical member length.

Problem 3.28. Find the displacements and internal forces on the truss structure shown. Assume that member 1 is given a temperature rise of T_0 and member 2 has a temperature distribution given by,

$$T = T_0\frac{y}{L}.$$

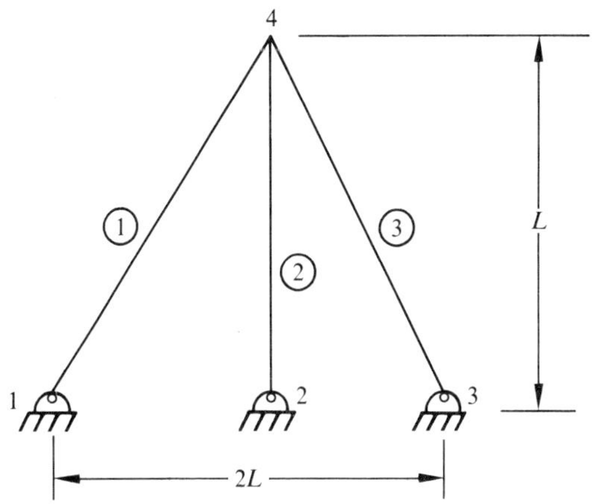

Problem 3.29. Given a tapered beam element as shown below. Derive its two-dimensional stiffness matrix equation. Assume a unit width.

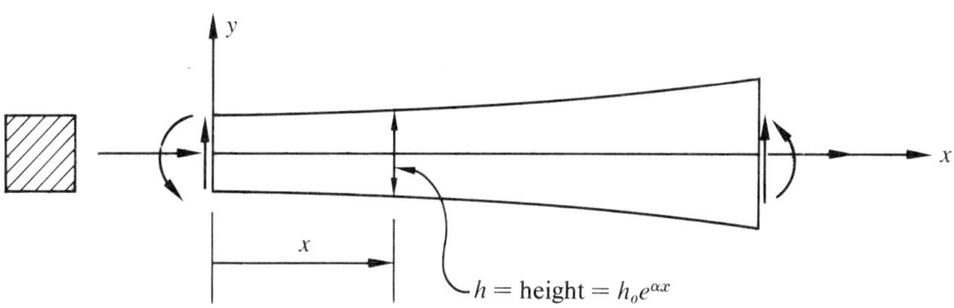

Problem 3.30. Express Eq. (3.51) in terms of the nodal coordinates of the curved beam element.

3.15 SUGGESTED COMPUTER EXERCISES

1. Write a computer program to solve for the displacements and internal loads for the truss-structure shown below. Assume the following to be your input data:

a. Number of nodes
b. Number of rods
c. Number of boundary conditions (zero restraints) and their locations
d. Number of external applied loads
e. Node coordinates
f. Nodes for each rod (high node, low node).
g. Properties for each rod (E, A)
h. External applied loads and their corresponding node location.

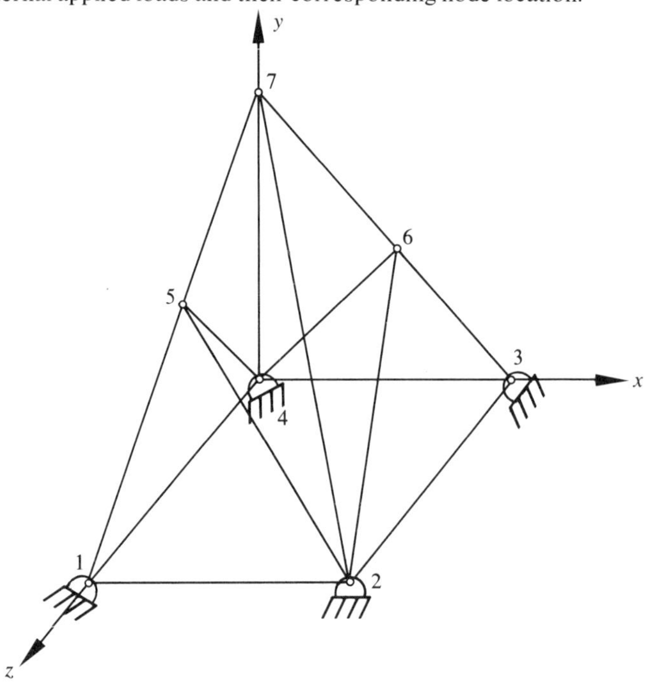

2. Write a general purpose program to compute the displacements and internal loads on a structure which consists of rectilinear beam elements. Assume the following to be your input data:

a. Number of nodes
b. Number of beams
c. Number of boundary conditions and their locations
d. Number of external applied loads
e. Node coordinates
f. Nodes for each beam (high node, low node)
g. Properties for each beam (E, I_y, I_z, A, J, G)
h. External applied loads and their corresponding node location.

CHAPTER FOUR

Element Stiffness Matrices of Lifting Surface Type Structures

4.1 INTRODUCTION

Missile fins, aircraft wings and tail surfaces are typical examples of lifting surface-type structures. The basic constituents of such structures are cover sheets (commonly referred to as the skin), reinforced with stiffeners and stabilized with spars and ribs. Figure 4.1 shows an exploded view of such surfaces.

Spars and rib elements are usually defined as beams of relatively heavy flanges separated by thin webs. In contrast to the slender solid beams, the shear deformation contribution to spar and rib deflections are significant. Therefore, it is essential to include shear web flexibility effects in the design of such beams.

Skin elements are basically considered as relatively thin plates when their force carrying capacity is in shear only. Thus, they will be referred to as shear panels. The skin stiffeners are defined as axial rods, and are specificly designed to resist normal stresses which cannot be accounted for by the skin. Hence, in this chapter the derivation and the application of the stiffness matrices of the following elements are presented: Spar or Rib Element, Rectangular Shear Panel Element, Trapezium Shear Panel Element, Triangular Membrane Element (In-Plane Forces only), Rectangular Membrane Element (In-Plane Forces only).

4.2 SPAR OR RIB ELEMENT STIFFNESS MATRIX

Figure 4.2 shows a spar or rib element which consists of heavy flanges separated by a thin web. The coordinate forces and displacements are positive as shown. The geometrical and mechanical properties of the

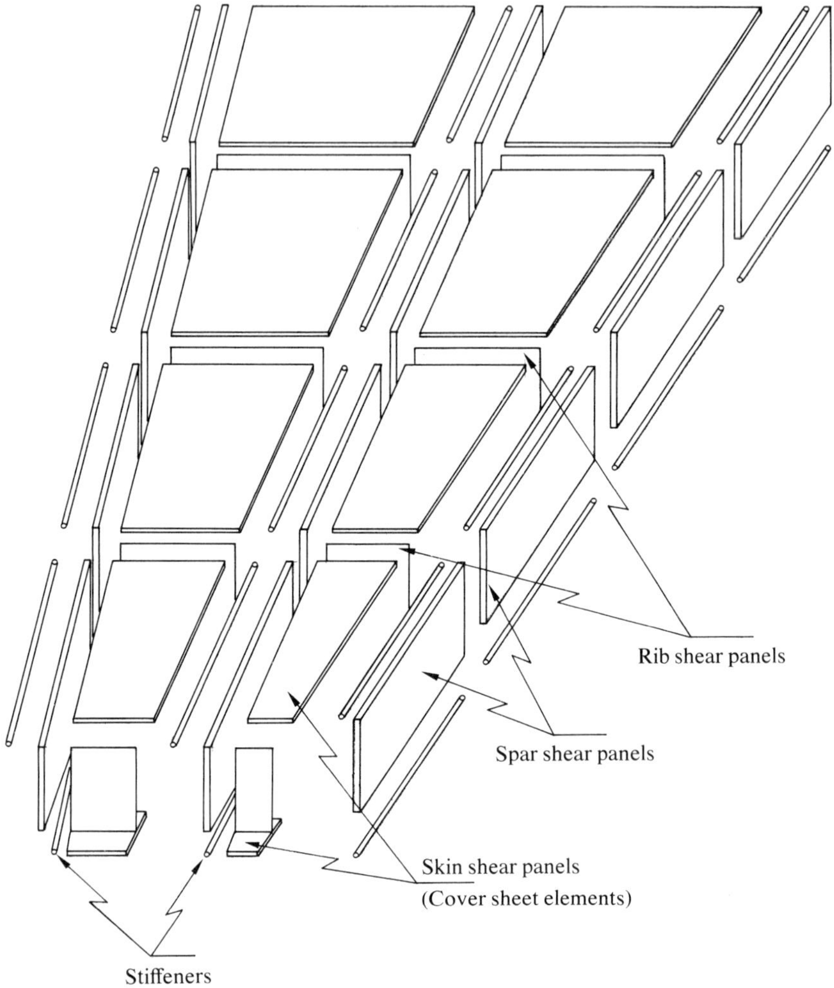

Fig. 4.1. Idealized wing, tail or fin structure.

element are defined as follows:

E_f = Young's Modulus of flange
G_w = shear modulus of web
L_y = depth of beam
L_x = length of beam
t_w = web thickness
A_w = flange area = $t_w L_y$

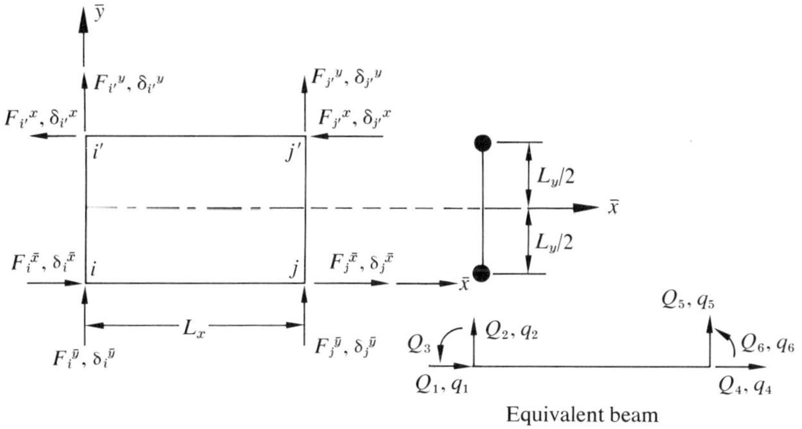

Fig. 4.2. Rib or spar element.

The following hypotheses are essential in deriving the stiffness matrix of the element shown in Fig. 4.2:

H.1. The stiffness in the $\bar{z}$ direction is negligible.

H.2. The flanges carry normal stresses only.

H.3. The web carries shearing stresses only.

H.4. Elementary beam theory is assumed, i.e.,

$$\delta_i^{\bar{x}} = -\delta_{i'}^{\bar{x}}, \qquad \delta_i^{\bar{y}} = \delta_{i'}^{\bar{y}}, \qquad \delta_j^{\bar{x}} = -\delta_{j'}^{\bar{x}}, \qquad \delta_j^{\bar{y}} = \delta_{j'}^{\bar{y}}.$$

It is necessary that the following substitutions be made,

$$\left.\begin{aligned} F_i^{\bar{x}} L_y &= Q_1 \\ F_j^{\bar{x}} L_y &= Q_3 \end{aligned}\right\} \text{equivalent bending moments}$$

$$\left.\begin{aligned} 2F_i^{\bar{y}} &= Q_2 \\ 2F_j^{\bar{y}} &= Q_4 \end{aligned}\right\} \text{equivalent shears}$$

$$\left.\begin{aligned} 2\delta_i^{x}/L_y &= q_1 \\ 2\delta_j^{x}/L_y &= q_3 \end{aligned}\right\} \text{equivalent rotation} \tag{4.1a}$$

$$\left.\begin{aligned} \delta_i^{\bar{y}} &= q_2 \\ \delta_j^{\bar{y}} &= q_4 \end{aligned}\right\} \text{equivalent translation}$$

The strain energy expression in bending and shear is given by,

$$U = \tfrac{1}{2}\int_0^{L_x} \frac{M^2}{EI}\,dx + \tfrac{1}{2}\int_0^{L_x} \frac{V^2}{A_w G_w}\,dx, \tag{4.1b}$$

or

$$q_i = \frac{\partial U}{\partial Q_i} = \int_0^{Ix} \frac{M}{EI}\frac{\partial M}{\partial Q_i}\,dx + \int_0^{Lx} \frac{V}{A_w G_w}\frac{\partial V}{\partial Q_i}\,dx \tag{4.1c}$$

Where M and V in terms of coordinate forces at node i are given by,

$$\begin{aligned} M &= Q_2 x - Q_1. \\ V &= -Q_2. \end{aligned} \tag{4.1d}$$

Utilizing Eq. (4.1c) and (4.1d), expanding for Q_i $(i = 1, 2)$ and integrating yields,

$$\begin{aligned} q_1 &= Q_1 L_x/EI - Q_2 L_x^2/2EI, \\ q_2 &= Q_2 L_x^3/3EI - Q_1 L_x^2/2EI + Q_2 L_x/A_w G_w, \end{aligned} \tag{4.1e}$$

in which q_1 and q_2 are the slope and vertical displacement at node i, respectively.

In matrix form, Eq. (4.1e) becomes,

$$\begin{bmatrix} q_1 \\ \\ q_2 \end{bmatrix} = \begin{bmatrix} \dfrac{L_x}{EI} & -\dfrac{L_x^2}{2EI} \\ \\ -\dfrac{L_x^2}{2EI} & \dfrac{L_x^3}{3EI} + \dfrac{L_x}{A_w G_w} \end{bmatrix} \begin{bmatrix} Q_1 \\ \\ Q_2 \end{bmatrix} \tag{4.1f}$$

The element stiffness coefficients will be obtained by the same procedure as presented in Chapter 3. Thus, if we let $q_1 = 1$ and all other q's in Eq. (4.1f) equal zero, yields;

$$\frac{Q_1}{q_1} = \bar{S}_{11} = \frac{4EI}{L_x} \frac{(1+\varphi)}{(1+4\varphi)}$$

and

$$\frac{Q_2}{q_1} = \bar{S}_{21} = \frac{6EI}{L_x^2} \frac{(1)}{1+4\varphi},$$

where

$$\varphi = 3EI/A_w G_w L_x^2.$$

From equilibrium;

$$Q_3 = -Q_1 + Q_2 L_x = \bar{S}_{31} = \frac{2EI}{L_x} \left(\frac{1-2\varphi}{1+4\varphi} \right),$$

$$Q_4 = -Q_2 = \bar{S}_{41} = -\frac{6EI}{L_x^2} \frac{1}{(1+4\varphi)}.$$

Now, if we let $q_2 = 1$ and all other q's equal zero, and repeating the above procedure, the following is obtained,

$$\frac{Q_1}{q_2} = \bar{S}_{12} = \frac{6EI}{L_x^2} \frac{1}{1+4\varphi},$$

$$\frac{Q_2}{q_2} = \bar{S}_{22} = \frac{12EI}{L_x^2} \frac{1}{1+4\varphi},$$

$$\frac{Q_3}{q_2} = \bar{S}_{32} = \frac{6EI}{L_x^2}\frac{1}{1+4\varphi},$$

$$\frac{Q_4}{q_2} = \bar{S}_{42} = \frac{12EI}{L_x^2}\frac{1}{1+4\varphi}.$$

If Eq. (4.1d) is expressed in terms of Q_3 and Q_4, and following the same steps as above, we obtain the following,

$$\frac{Q_3}{q_3} = \bar{S}_{33} = \frac{4EI}{L_x}\frac{(1+\varphi)}{(1+4\varphi)}$$

$$\frac{Q_4}{q_3} = \bar{S}_{43} = -\frac{6EI}{L_x^2}\frac{1}{1+4\varphi}$$

$$S_i = \bar{S}_{3i} \qquad (i = 1, 2)$$

$$\frac{Q_4}{q_4} = \bar{S}_{44} = \frac{12EI}{L_x^2}\frac{1}{1+4\varphi}$$

$$S_{i4} = \bar{S}_{4i} \qquad (i, = 1, 2, 3)$$

Having calculated the stiffness coefficients, we can now convert to our common nomenclature (forces and displacements) as given in Eq. (4.1a). Thus, making the conversion, and letting $n = L_y/L_x$, yields the following stiffness matrix equation,

$$\begin{bmatrix} F_i^{\bar{x}} \\ F_i^{\bar{y}} \\ F_j^{\bar{x}} \\ F_j^{\bar{y}} \end{bmatrix} = \frac{6EI}{(1+4\varphi)L_xL_y^2}\begin{bmatrix} \frac{4}{3}(1+\varphi) & & \text{Sym.} & \\ n & n^2 & & \\ \frac{2}{3}(1-2\varphi) & n & \frac{4}{3}(1+\varphi) & \\ -n & -n^2 & -n & n^2 \end{bmatrix}\begin{bmatrix} \delta_i^{\bar{x}} \\ \delta_i^{\bar{y}} \\ \delta_j^{\bar{x}} \\ \delta_j^{\bar{y}} \end{bmatrix} \tag{4.2}$$

or in shorthand matrix notation, Eq. (4.2) becomes,

$$\{\bar{Q}\} = [\bar{S}]\{\bar{q}\}, \tag{4.2a}$$

where $[\bar{S}]$ is defined in Eq. (4.2).

To make Eq. (4.2) applicable to an element which is arbitrary oriented, we use the transformation matrix derived previously (*see* Eq. 3.15).

Thus Eq. (4.2) in the reference coordinate system becomes,

$$\begin{bmatrix} F_i^x \\ F_i^y \\ F_i^z \\ F_j^x \\ F_j^y \\ F_j^z \end{bmatrix} = \begin{bmatrix} [T] & \\ & [T] \end{bmatrix}^{-1} [\bar{S}] \begin{bmatrix} [T] & \\ & [T] \end{bmatrix} \begin{bmatrix} \delta_i^x \\ \delta_i^y \\ \delta_i^z \\ \delta_j^x \\ \delta_j^y \\ \delta_j^z \end{bmatrix} \tag{4.3}$$

4.3 RECTANGULAR SHEAR PANEL ELEMENT STIFFNESS MATRIX

Shown in Fig. 4.3 is a shear panel element of a constant thickness t with uniform continuous shear flow τ_0 acting along the edges of the panel.

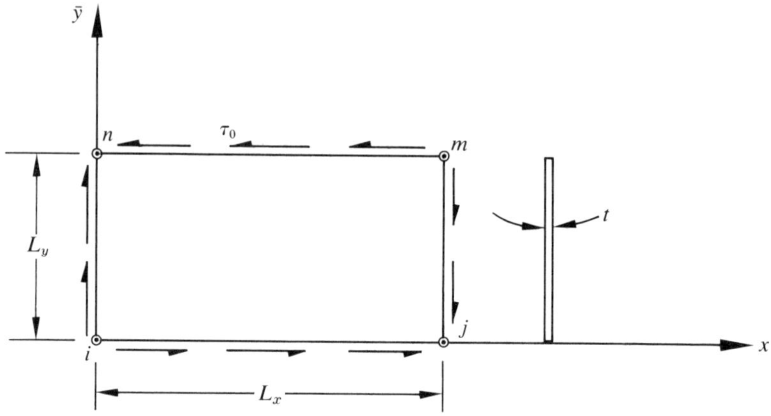

Fig. 4.3. Shear panel element with continuous uniform shear flow.

Figure 4.4 shows the same element with the shear flow being replaced by lumped forces at the nodes (corners in this case) of the panel, i.e.,

$$\begin{aligned} F_i^{\bar{x}} &= F_j^{\bar{x}} = \tau_0 L_x/2 \\ F_m^{\bar{x}} &= F_n^{\bar{x}} = -\tau_0 L_x/2 \\ F_j^{\bar{y}} &= F_m^{\bar{y}} = \tau_0 L_y/2 \end{aligned} \tag{4.4a}$$

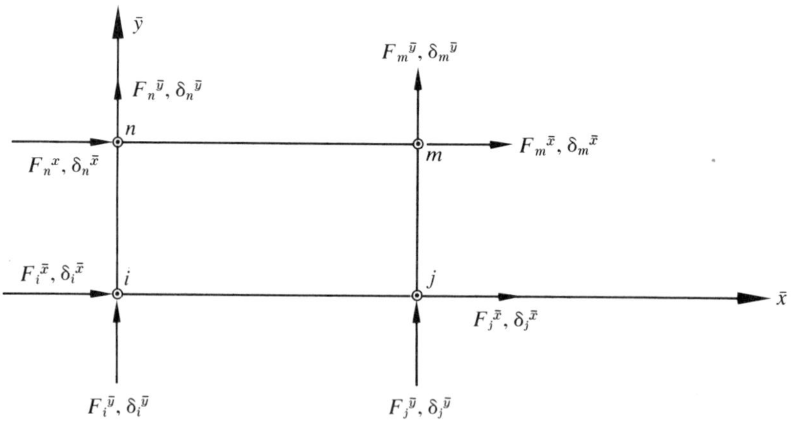

Fig. 4.4. Shear panel element with lumped forces.

To derive the stiffness coefficients, Eq. (3.3) will be used, i.e.,

$$S_{ij} = \frac{\partial^2 U}{\partial q_i \partial q_j} \tag{3.3}$$

where U must be expressed in terms of the coordinate displacements.

From Eq. (2.31) where shearing strain only is considered, the strain energy is given by,

$$U = \tfrac{1}{2} \int_0^{L_y} \int_0^{L_x} \int_{-t/2}^{t/2} G \epsilon_{\bar{x}\bar{y}}^2 \, dt \, d\bar{x} \, d\bar{y}. \tag{4.4b}$$

The coordinate displacements are defined in terms of the generalized displacement notation as follows,

$$\begin{bmatrix} q_1 \\ q_2 \\ q_3 \\ q_4 \\ q_5 \\ q_6 \\ q_7 \\ q_8 \end{bmatrix} = \begin{bmatrix} \delta_i^{\bar{x}} \\ \delta_i^{\bar{y}} \\ \delta_j^{\bar{x}} \\ \delta_j^{\bar{y}} \\ \delta_m^{\bar{x}} \\ \delta_m^{\bar{y}} \\ \delta_n^{\bar{x}} \\ \delta_n^{\bar{y}} \end{bmatrix} \tag{4.4c}$$

Also, let q_x and q_y be the displacements at any point in the $\bar{x}$- and $\bar{y}$-directions, respectively.

From Eq. (2.28) and dropping the bars for convenience, the shearing strain is defined as,

$$\epsilon_{xy} = \frac{\partial q_x}{\partial y} + \frac{\partial q_y}{\partial x}. \tag{4.4d}$$

The problem, now, is to be able to obtain expressions for q_x and q_y in terms of the coordinate nodal displacements q_i ($i = 1$–8). This may be accomplished by adopting the hypothesis that the shearing strain is a constant, throughout the panel, i.e.,

$$\epsilon_{xy} = \frac{\partial q_x}{\partial y} + \frac{\partial q_y}{\partial x} = C_1. \tag{4.4e}$$

From Eq. (4.4e) it can be easily deduced that the normal strain must be of the form,

$$\epsilon_{xx} = \frac{\partial q_x}{\partial x} = C_2 + C_3 y \tag{4.4f}$$

$$\epsilon_{yy} = \frac{\partial q_y}{\partial y} = C_4 + C_5 x \tag{4.4g}$$

where C_i $(i = 1, 2, \ldots, 5)$ are constants to be determined from boundary conditions.

Upon integrating Eqs. (4.4f) and (4.4g) the following is obtained,

$$q_x = C_2 x + C_3 yx + F_1(y), \tag{4.4h}$$

$$q_y = C_4 y + C_5 xy + F_2(x), \tag{4.4i}$$

where $F_1(y)$ and $F_2(x)$ are constants of integration.

Equations (4.4h) and (4.4i) substituted into Eq. (4.4e) will yield,

$$C_3 x + \frac{dF_2}{dx} + C_5 y + \frac{dF_1}{dy} = C_1 \tag{4.4j}$$

or

$$G_1(x) + G_2(y) = C_1, \tag{4.4k}$$

where

$$G_1(x) = C_3 x + \frac{dF_2}{dx}, \tag{4.4l}$$

$$G_2(y) = C_5 - y + \frac{dF_1}{dy}. \tag{4.4m}$$

Note that $G_1(x)$ and $G_2(y)$ are two independent functions, and for their sum to be a constant (*see* Eq. 4.4k) each function must be a constant itself, i.e.,

$$G_2(y) = \frac{dF_1}{dy} + C_5 y = C_6, \tag{4.4n}$$

and

$$G_1(x) = \frac{dF_2}{dx} + C_3 x = C_9. \tag{4.4o}$$

Equations (4.4n) and (4.4o) are two linear ordinary differential equations whose solutions may readily be obtained by direct integration,

$$F_1 = -\frac{C_5}{2} y^2 + C_6 y + C_7, \tag{4.4p}$$

$$F_2 = -\frac{C_3}{2} x^2 + C_9 x + C_8. \tag{4.4q}$$

Substituting Eqs. (4.4p, q) into Eqs. (4.4h, i) yield,

$$\begin{aligned} q_x(x, y) &= C_2 x + C_3 xy + C_6 y + C_7 - C_5 y^2/2, \\ q_y(x, y) &= C_4 y + C_5 xy + C_9 x + C_8 - C_3 x^2/2. \end{aligned} \tag{4.5}$$

Note that,

$$C_9 = C_1 - C_6. \tag{4.5a}$$

Therefore Eq. (4.5b) becomes,

$$q_y(x, y) = C_4 y + C_5 xy + (C_1 - C_6)x + C_8 - C_3 x^2/2. \tag{4.5b}$$

To determine the constants, C_i $(i = 1, 2, \ldots, 8)$ in terms of the nodal coordinate displacements q_i $(i = 1, 2, \ldots, 8)$, the following boundary conditions are imposed (Fig. 4.4),

1. $q_x(0, 0) = q_1 = C_7$

   $q_y(0, 0) = q_2 = C_8$

2. $q_x(L_x, 0) = q_3 = L_x C_2 + C_7$

   $q_y(L_x, 0) = q_4 = (C_1 - C_6)L_x + C_8 - \dfrac{L_x^2}{2} C_3$

3. $q_x(L_x, L_y) = q_5 = L_x C_2 + L_x L_y C_3 + L_y C_6 + C_7 - \dfrac{L_y^2}{2} C_5$

   $q_y(L_x, L_y) = q_6 = L_y C_4 + L_x L_y C_5 + (C_1 - C_6)L_x + C_8 - \dfrac{L_x^2}{2} C_3$

4. $q_x(0, L_y) = q_7 = L_y C_6 + C_7 - \dfrac{L_y^2}{2} C_5$

   $q_y(0, L_y) = q_8 = L_y C_4 + C_8$

Solving the above equations for C_i $(i = 1, 2, \ldots, 8)$ and substituting back into Eq. (4.5) yield,

$$\begin{aligned} q_x &= (q_3 - q_1)x/L_x + (q_5 - q_7 + q_1 - q_3)xy/L_x L_y + [(q_7 - q_1)/L_y \\ &\quad + (q_6 + q_2 - q_4 - q_8)/2L_x]y + q_7 - (q_6 + q_2 - q_4 - q_8)y^2/2L_x L_y \\ q_y &= (q_8 - q_2)y/L_y + (q_6 + q_2 - q_4 - q_8)xy/L_x L_y + [(q_4 - q_2)/L_x \\ &\quad + (q_1 + q_5 - q_7 - q_3)/2L_y]x + q_2 - (q_5 - q_7 + q_1 - q_3)x^2/2L_x L_y \end{aligned} \tag{4.6}$$

Hence the sought shearing strain in terms of nodal coordinate displacements becomes:

$$\epsilon_{xy} = (q_5 + q_7 - q_1 - q_3)/2L_y + (q_6 + q_4 - q_2 - q_8)/2L_x. \tag{4.6a}$$

Substituting Eq. (4.6a) into Eq. (4.4b) utilizing Eq. (3.3) and integrating yields the following stiffness matrix for the rectangular shear panel element.

Where

$$r = L_x/L_y,$$

$$[\bar{S}] = \frac{Gt}{4}\begin{bmatrix} r & & & & & & & \\ 1 & 1/r & & & \text{Sym.} & & & \\ r & 1 & r & & & & & \\ -1 & -r & -1 & 1/r & & & & \\ -r & -1 & -r & 1 & r & & & \\ -1 & -1/r & -1 & 1/r & 1 & 1/r & & \\ -r & -1 & -r & 1 & r & 1 & r & \\ 1 & 1/r & 1 & -1/r & -1 & -1/r & -1 & 1/r \end{bmatrix} \tag{4.7}$$

The stiffness matrix equation can be written as,

$$\begin{bmatrix} F_i^{\bar{x}} \\ F_i^{\bar{y}} \\ F_j^{\bar{x}} \\ F_j^{\bar{y}} \\ F_m^{\bar{x}} \\ F_m^{\bar{y}} \\ F_n^{\bar{x}} \\ F_n^{\bar{y}} \end{bmatrix} = [\bar{S}] \begin{bmatrix} \delta_i^{\bar{x}} \\ \delta_i^{\bar{y}} \\ \delta_j^{\bar{x}} \\ \delta_j^{\bar{y}} \\ \delta_m^{\bar{x}} \\ \delta_m^{\bar{y}} \\ \delta_n^{\bar{x}} \\ \delta_n^{\bar{y}} \end{bmatrix} \tag{4.8}$$

where $[\bar{S}]$ is defined in Eq. (4.7).

Here again Eq. (4.8) is applicable only to elements oriented as shown in Fig. 4.3. To render its application to elements, arbitrarily oriented, we shall develop a transformation matrix which makes such application possible.

The transformation will be carried based on the trapezium element, of which the rectangular element is a special case. This transformation is accomplished in two configurations of element orientation (*see* Fig. 4.5).

Case 1.

1. Rotation about the y-axis.
2. Subsequent rotation about the z_1-axis.

Case 2.

1. Rotation about the y-axis.
2. Subsequent rotation about the x_1-axis.

Case 1 is identical to that of the straight beam. Hence, from Eq. (3.9) the transformation matrix is,

$$[T]_{\text{case 1}} = \begin{bmatrix} \cos\alpha\cos\beta & \sin\beta & \cos\beta\sin\alpha \\ -\cos\alpha\sin\beta & \cos\beta & -\sin\alpha\sin\beta \\ -\sin\alpha & 0 & \cos\alpha \end{bmatrix} \tag{4.9}$$

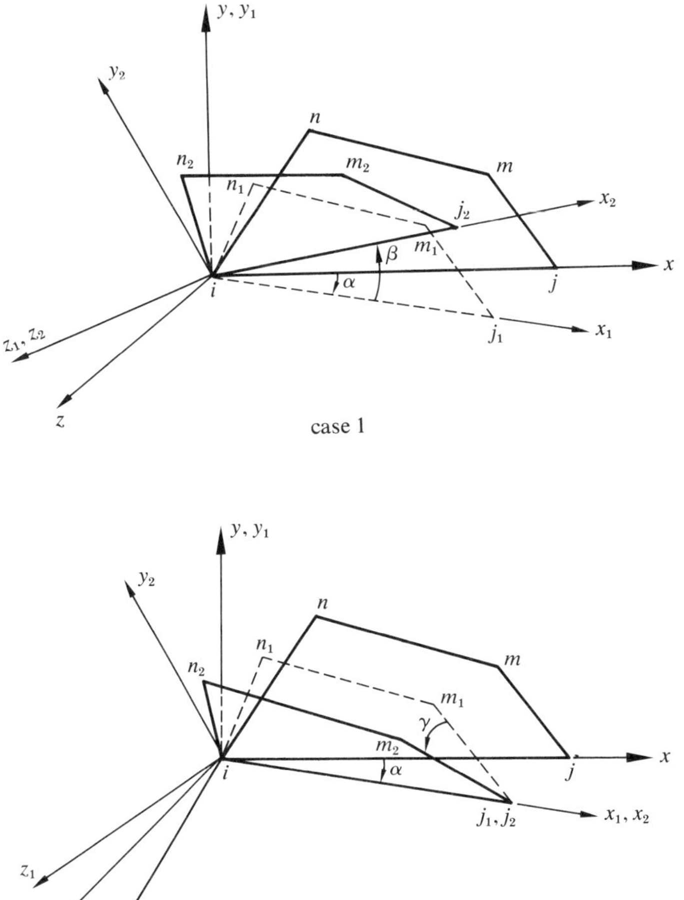

Fig. 4.5. Element rotations.

In terms of nodal coordinates, Eq. (4.9) becomes,

$$[T] = \begin{bmatrix} \dfrac{x_{ji}}{L} & \dfrac{y_{ji}}{L} & \dfrac{z_{ji}}{L} \\ -\dfrac{x_{ji}y_{ji}}{L\mathscr{S}} & \dfrac{\mathscr{S}}{L} & -\dfrac{y_{ji}z_{ji}}{L\mathscr{S}} \\ -\dfrac{z_{ji}}{\mathscr{S}} & 0 & \dfrac{x_{ji}}{L} \end{bmatrix} \tag{4.9a}$$

For elements which are oriented along the y-axis ($x_{ji} = z_{ji} = 0$), Eq. (4.7a) reduces to,

$$[T] = \begin{bmatrix} 0 & 1 & 0 \\ -1 & 0 & 0 \\ 0 & 0 & 1 \end{bmatrix} \qquad (\text{if } y_{ji} \text{ is positive}) \tag{4.9b}$$

or

$$[T] = \begin{bmatrix} 0 & -1 & 0 \\ 1 & 0 & 0 \\ 0 & 0 & 1 \end{bmatrix} \qquad (\text{if } y_{ji} \text{ is negative}) \tag{4.9c}$$

$$x_{ji} = x_j - x_i, \text{ etc.}$$

$$L = (x_{ji}^2 + y_{ji}^2 + z_{ji}^2)^{1/2}, \tag{4.9d}$$

and

$$\mathscr{S} = (x_{ji}^2 + z_{ji}^2)^{1/2}.$$

In an analogous manner to the derivation of Eq. (3.9) the transformation matrix for Case 2 can be easily obtained and is given as,

$$[T]_{\text{case 2}} = \begin{bmatrix} \cos\alpha & 0 & \sin\alpha \\ -\sin\alpha\sin\alpha & \cos\alpha & \cos\alpha\sin\alpha \\ -\sin\alpha\cos\alpha & -\sin\alpha & \cos\alpha\cos\alpha \end{bmatrix} \tag{4.10}$$

In terms of notal coordinates, Eq. (4.8) becomes,

$$[T] = \begin{bmatrix} \dfrac{x_{ji}}{\rho_1} & 0 & \dfrac{z_{ji}}{\rho_1} \\[2ex] \dfrac{z_{ji}\rho_3}{\rho_1\rho_2} & \dfrac{y_{ni}}{\rho_2} & \dfrac{x_{ji}\rho_3}{\rho_1\rho_2} \\[2ex] -\dfrac{z_{ji}y_{ni}}{\rho_1\rho_2} & -\dfrac{\rho_3}{\rho_2} & \dfrac{x_{ji}y_{ni}}{\rho_1\rho_2} \end{bmatrix} \tag{4.10a}$$

where

$$\begin{aligned} \rho_1 &= (x_{ji}^2 + z_{ji}^2)^{1/2} \\ \rho_2 &= [(x_{ni} - \Delta_3)^2 + (z_{ni} - \Delta_4)^2 + y_{ni}^2]^{1/2} \\ \Delta_3 &= (\Delta_5 + \Delta_6{}^2 + \rho_1{}^2)(x_{ji})/(2\rho_1{}^2) \\ \Delta_4 &= \Delta_3 z_{ji}/x_{ji} \\ \Delta_5 &= x_{nj}^2 + y_{nj}^2 + z_{nj}^2 \end{aligned} \tag{4.10b}$$

$$\Delta_6 = (x_{ni}^2 + y_{ni}^2 + z_{ni}^2)^{1/2}$$

$$\rho_3 = (\rho_2{}^2 - y_{ni}^2)^{1/2}$$

$$x_{ji} = x_j - x_i,\ x_{ni} = x_n - x_i,\ \text{etc.}$$

Hence, Eq. (4.8) in the reference coordinate system becomes;

$$\begin{bmatrix} F_i^x \\ F_i^y \\ F_i^z \\ F_j^x \\ F_j^y \\ F_j^z \\ F_m^x \\ F_m^y \\ F_m^z \\ F_n^x \\ F_n^y \\ F_n^z \end{bmatrix} = \begin{bmatrix} [T] & & & \\ & [T] & & \\ & & [T] & \\ & & & [T] \end{bmatrix}^{-1} [\bar{S}] \begin{bmatrix} [T] & & & \\ & [T] & & \\ & & [T] & \\ & & & [T] \end{bmatrix} \begin{bmatrix} \delta_i^x \\ \delta_i^y \\ \delta_i^z \\ \delta_j^x \\ \delta_j^y \\ \delta_j^z \\ \delta_m^x \\ \delta_m^y \\ \delta_m^z \\ \delta_n^x \\ \delta_n^y \\ \delta_n^z \end{bmatrix} \tag{4.11}$$

or in shorthand matrix notation,

$$\{Q\} = [S]\{q\} \tag{4.11a}$$

where

$$[S] = \begin{bmatrix} [T] & & & \\ & [T] & & \\ & & [T] & \\ & & & [T] \end{bmatrix}^{-1} [\bar{S}] \begin{bmatrix} [T] & & & \\ & [T] & & \\ & & [T] & \\ & & & [T] \end{bmatrix} \tag{4.11b}$$

The transformation matrix $[T]$ is given in Eq. (4.9) or (4.10) depending on the element orientation; $[\bar{S}]$ is defined in Eq. (4.7) with the exception of adding rows and columns of zeros corresponding to $[F_r^z;\ \delta_r^z\ (r = i, j, m, n)]$.

4.4 TRAPEZIUM SHEAR PANEL ELEMENT STIFFNESS MATRIX

The stiffness coefficients for the trapezium shear panel can be derived in an analogous manner to those of the rectangular shear panel. Hence, excluding the detailed long derivations, the stiffness matrix equation in local coordinates is given by,

$$\begin{bmatrix} F_i^{\bar{x}} \\ F_i^{\bar{y}} \\ F_j^{\bar{x}} \\ F_j^{\bar{y}} \\ F_m^{\bar{x}} \\ F_m^{\bar{y}} \\ F_n^{\bar{x}} \\ F_n^{\bar{y}} \end{bmatrix} = \frac{tG}{2A} \begin{bmatrix} x_{nj}^2 & & & & & & & \\ -y_{ni}x_{nj} & y_{ni}^2 & & & & \text{Sym.} & & \\ -x_{mi}x_{nj} & y_{ni}x_{mj} & x_{mi}^2 & & & & & \\ y_{mi}x_{nj} & -y_{mi}y_{ni} & -y_{mi}x_{mi} & y_{mi}^2 & & & & \\ -x_{nj}^2 & y_{ni}x_{nj} & x_{mi}x_{nj} & -y_{mi}x_{nj} & x_{nj}^2 & & & \\ y_{ni}x_{nj} & -y_{ni}^2 & -y_{ni}x_{mi} & y_{mi}y_{ni} & -y_{ni}x_{nj} & y_{ni}^2 & & \\ x_{mi}x_{nj} & -y_{ni}x_{mi} & -x_{mi}^2 & y_{mi}x_{mi} & -x_{mi}x_{nj} & y_{ni}x_{mt} & x_{mt}^2 & \\ -y_{mi}x_{nj} & y_{mi}y_{ni} & y_{mi}x_{mi} & -y_{mi}^2 & y_{mi}x_{nj} & -y_{mi}y_{ni} & -y_{mi}x_{mi} & y_{mi}^2 \end{bmatrix} \begin{bmatrix} \delta_i^{\bar{x}} \\ \delta_i^{\bar{y}} \\ \delta_j^{\bar{x}} \\ \delta_j^{\bar{y}} \\ \delta_m^{\bar{x}} \\ \delta_m^{\bar{y}} \\ \delta_n^{\bar{x}} \\ \delta_n^{\bar{y}} \end{bmatrix} \tag{4.12}$$

where $x_{nj} = x_n - x_j$, etc.

$$A = (x_{ji}y_{mi} + x_{mi}y_{nj} - x_{ni}y_{mi}) \tag{4.12a}$$

$t =$ panel thickness.

Equation (4.12) is applicable only to elements oriented as shown in Fig. 4.6. To render its application to arbitrarily oriented elements we employ the transformation technique utilized previously. Thus, in the reference coordinate system Eq. (4.12) becomes,

$$\begin{bmatrix} F_i^x \\ F_i^y \\ F_i^z \\ F_j^x \\ F_j^y \\ F_j^z \\ F_m^x \\ F_m^y \\ F_m^z \\ F_n^x \\ F_n^y \\ F_n^z \end{bmatrix} = \begin{bmatrix} [T] & & & \\ & [T] & & \\ & & [T] & \\ & & & [T] \end{bmatrix}^{-1} [\bar{S}] \begin{bmatrix} [T] & & & \\ & [T] & & \\ & & [T] & \\ & & & [T] \end{bmatrix} \begin{bmatrix} \delta_i^x \\ \delta_i^y \\ \delta_i^z \\ \delta_j^x \\ \delta_j^y \\ \delta_j^z \\ \delta_m^x \\ \delta_m^y \\ \delta_m^z \\ \delta_n^x \\ \delta_n^y \\ \delta_n^z \end{bmatrix} \tag{4.13}$$

or, in shorthand matrix notation, Eq. (4.13) becomes,

$$\{Q\} = [S]\{q\}, \tag{4.13a}$$

where

$$[S] = [T]^{-1}[\bar{S}][T]. \tag{4.13b}$$

The coefficients of the transformation $[T]$ are given in Eq. (4.9) or (4.10) depending on the element rotation; and the stiffness matrix $[\bar{S}]$ is defined in Eq. (4.12) with the exception of adding rows and columns of zeros corresponding to $[F_r^z; \delta_r^z \ (r = i, j, m, n)]$.

Upon carrying out the matrix multiplication in Eq. (4.13b) the following

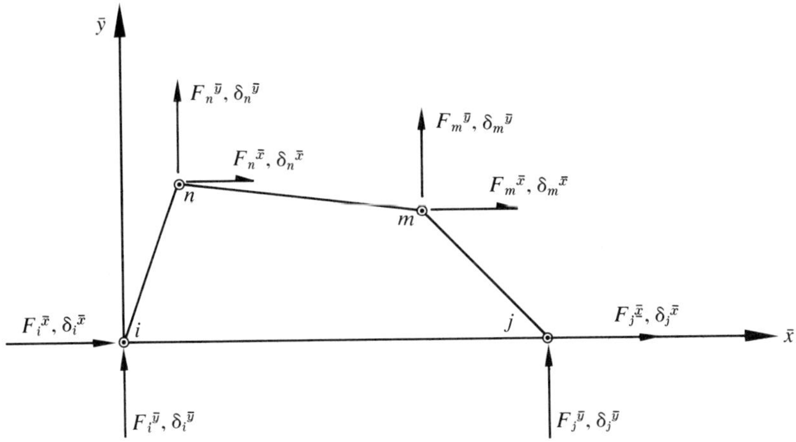

Fig. 4.6. Trapezium shear panel.

generalized stiffness coefficients for a trapezium structural shear element are obtained,

$$
\begin{aligned}
&\bar{S}_{11} = T_{11}^2\mathscr{S}_1 + 2T_{11}T_{21}\mathscr{S}_2 + T_{21}^2\mathscr{S}_3 \\
&\bar{S}_{12} = T_{11}T_{12}\mathscr{S}_1 + T_{12}T_{21}\mathscr{S}_2 + T_{22}T_{11}\mathscr{S}_2 + T_{22}T_{21}\mathscr{S}_3 \\
&\bar{S}_{13} = T_{13}T_{11}\mathscr{S}_1 + T_{13}T_{21}\mathscr{S}_2 + T_{23}T_{11}\mathscr{S}_2 + T_{23}T_{21}\mathscr{S}_3 \\
&\bar{S}_{14} = T_{11}^2\mathscr{S}_4 + T_{11}T_{21}\mathscr{S}_5 + T_{21}T_{11}\mathscr{S}_7 + T_{22}^2\mathscr{S}_8 \\
&\bar{S}_{15} = T_{12}T_{11}\mathscr{S}_4 + T_{12}T_{21}\mathscr{S}_5 + T_{22}T_{11}\mathscr{S}_7 + T_{22}T_{21}\mathscr{S}_8 \\
&\bar{S}_{16} = T_{13}T_{11}\mathscr{S}_4 + T_{13}T_{21}\mathscr{S}_5 + T_{23}T_{11}\mathscr{S}_7 + T_{23}T_{21}\mathscr{S}_8 \\
&\bar{S}_{1r}\ (r = 7, 8, \ldots, 12) = -\bar{S}_{1j}\ (j = 1, 2, \ldots, 6) \\
&\bar{S}_{21} = \bar{S}_{12} \\
&\bar{S}_{22} = T_{12}^2\mathscr{S}_1 + 2T_{12}T_{22}\mathscr{S}_2 + T_{22}^2\mathscr{S}_3 \\
&\bar{S}_{23} = T_{13}T_{12}\mathscr{S}_1 + T_{13}T_{22}\mathscr{S}_2 + T_{23}T_{12}\mathscr{S}_2 + T_{23}T_{22}\mathscr{S}_3 \\
&\bar{S}_{24} = T_{11}T_{12}\mathscr{S}_4 + T_{11}T_{22}\mathscr{S}_5 + T_{21}T_{12}\mathscr{S}_7 + T_{21}T_{22}\mathscr{S}_8 \\
&\bar{S}_{25} = T_{12}^2\mathscr{S}_4 + T_{12}T_{22}\mathscr{S}_5 + T_{22}T_{12}\mathscr{S}_7 + T_{22}^2\mathscr{S}_8 \\
&\bar{S}_{26} = T_{13}T_{12}\mathscr{S}_4 + T_{13}T_{22}\mathscr{S}_5 + T_{23}T_{12}\mathscr{S}_7 + T_{23}T_{22}\mathscr{S}_8 \\
&\bar{S}_{2r}\ (r = 7, 8, \ldots, 12) = -\bar{S}_{2j}\ (j = 1, 2, \ldots, 6) \\
&\bar{S}_{31} = \bar{S}_{13} \\
&\bar{S}_{32} = \bar{S}_{23} \\
&\bar{S}_{33} = T_{13}^2\mathscr{S}_1 + 2T_{13}T_{23}\mathscr{S}_2 + T_{23}^2\mathscr{S}_3 \\
&\bar{S}_{34} = T_{11}T_{13}\mathscr{S}_4 + T_{11}T_{23}\mathscr{S}_5 + T_{21}T_{13}\mathscr{S}_7 + T_{21}T_{23}\mathscr{S}_8 \\
&\bar{S}_{35} = T_{12}T_{13}\mathscr{S}_4 + T_{12}T_{23}\mathscr{S}_5 + T_{22}T_{13}\mathscr{S}_7 + T_{22}T_{23}\mathscr{S}_8 \\
&\bar{S}_{36} = T_{13}^2\mathscr{S}_4 + T_{13}T_{23}\mathscr{S}_5 + T_{23}T_{13}\mathscr{S}_7 + T_{22}^2\mathscr{S}_8 \\
&\bar{S}_{3r}\ (r = 7, 8, \ldots, 12) = -\bar{S}_{3j}\ (j = 1, 2, \ldots, 6) \\
&\bar{S}_{4i} = S_{i4}\ (i = 1, 2, 3)
\end{aligned}
\qquad (4.13c)
$$

$$
\begin{aligned}
&\bar{S}_{44} = T_{11}^2\mathscr{S}_6 + 2T_{11}T_{21}\mathscr{S}_9 + T_{21}^2\mathscr{S}_{10} \\
&\bar{S}_{45} = T_{12}T_{11}\mathscr{S}_6 + T_{12}T_{21}\mathscr{S}_9 + T_{22}T_{11}\mathscr{S}_9 + T_{22}T_{21}\mathscr{S}_{10} \\
&\bar{S}_{46} = T_{13}T_{11}\mathscr{S}_6 + T_{13}T_{21}\mathscr{S}_9 + T_{23}T_{11}\mathscr{S}_9 + T_{23}T_{21}\mathscr{S}_{10} \\
&\bar{S}_{4r}\ (r = 7, 8, \ldots, 12) = -\bar{S}_{4j}\ (j = 1, 2, \ldots, 6) \\
&\bar{S}_{5i} = \bar{S}_{i5}\ (i = 1, 2, \ldots, 4) \\
&\bar{S}_{55} = T_{12}^2\mathscr{S}_6 + 2T_{12}T_{22}\mathscr{S}_9 + T_{22}^2\mathscr{S}_{10} \\
&\bar{S}_{56} = T_{13}T_{12}\mathscr{S}_6 + T_{13}T_{22}\mathscr{S}_9 + T_{23}T_{12}\mathscr{S}_9 + T_{23}T_{22}\mathscr{S}_{10} \\
&\bar{S}_{5r}\ (r = 7, 8, \ldots, 12) = -\bar{S}_{5j}\ (j = 1, 2, \ldots, 6) \\
&\bar{S}_{6i} = \bar{S}_{i6}\ (i = 1, 2, \ldots, 5) \\
&\bar{S}_{66} = T_{13}^2\mathscr{S}_6 + 2T_{13}T_{23}\mathscr{S}_9 + T_{23}^2\mathscr{S}_{10} \\
&\bar{S}_{6r}\ (r = 7, 8, \ldots, 12) = -\bar{S}_{6j}\ (j = 1, 2, \ldots, 6) \\
&\bar{S}_{7i} = \bar{S}_{i7}\ (i = 1, 2, \ldots, 6) \\
&\bar{S}_{7r}\ (r = 7, 8, \ldots, 12) = -\bar{S}_{7j}\ (j = 1, 2, \ldots, 6) \\
&\bar{S}_{8i} = \bar{S}_{i8}\ (i = 1, 2, \ldots, 7) \\
&\bar{S}_{8r}\ (r = 8, 9, \ldots, 12) = -\bar{S}_{8j}\ (j = 2, 3, \ldots, 6) \\
&\bar{S}_{9i} = \bar{S}_{i9}\ (i = 1, 2, \ldots, 8) \\
&\bar{S}_{9r}\ (r = 9, 10, \ldots, 12) = -\bar{S}_{9j}\ (j = 3, 4, \ldots, 6) \\
&\bar{S}_{10,r}\ (r = 10, 11, 12) = -\bar{S}_{10,j}\ (j = 4, 5, 6) \\
&\bar{S}_{11,i} = \bar{S}_{i,11}\ (i = 1, 2, \ldots, 10) \\
&\bar{S}_{11,r}\ (r = 11, 12) = -\bar{S}_{11,j}\ (j = 5, 6) \\
&\bar{S}_{12,i} = \bar{S}_{i,12}\ (i = 1, 2, \ldots, 11) \\
&\bar{S}_{12,12} = -\bar{S}_{12,6}
\end{aligned}
\tag{4.13c}
$$

in which

$$
\begin{aligned}
\mathscr{S}_1 &= \lambda_4(\lambda_3 - \lambda_2)^2 \\
\mathscr{S}_2 &= \lambda_4(\lambda_1 - \lambda_3)(\lambda_8) \\
\mathscr{S}_3 &= \lambda_4\lambda_8^2 \\
\mathscr{S}_4 &= \lambda_4\lambda_2(\lambda_1 - \lambda_3) \\
\mathscr{S}_5 &= \lambda_4\lambda_8\lambda_2 \\
\mathscr{S}_6 &= \lambda_4\lambda_2^2 \\
\mathscr{S}_7 &= \lambda_4\lambda_7(\lambda_3 - \lambda_1) \\
\mathscr{S}_8 &= -\lambda_4\lambda_7\lambda_8 \\
\mathscr{S}_9 &= -\lambda_4\lambda_7\lambda_2 \\
\mathscr{S}_{10} &= \lambda_4\lambda_7^2 \\
\lambda_1 &= \beta_1 \\
\lambda_2 &= (\beta_3^2 - \lambda_3^2)^{1/2} \\
\lambda_3 &= [ABS\ (\beta_7 - \lambda_8^2)]^{1/2} \qquad (ABS = \text{absolute value}) \\
\lambda_4 &= tG/[2(\lambda_1\lambda_7 + \lambda_2\lambda_8 - \lambda_3\lambda_7)] \\
\lambda_5 &= [(\beta_1^2 + \beta_7 - \beta_6)(x_{ji})/2\beta_1]^{1/2} \\
\lambda_6 &= (\lambda_5 z_{ji}/x_{ji})^{1/2} \\
\lambda_7 &= \beta_2\beta_5 \\
\lambda_8 &= [(x_{ni} - \lambda_5)^2 + (z_{ni} - \lambda_6)^2 + y_{ni}^2]^{1/2}
\end{aligned}
\tag{4.13d}
$$

$$\begin{aligned}
\beta_1 &= (x_{ji}^2 + y_{ji}^2 + z_{ji}^2)^{1/2} \\
\beta_2 &= (x_{mj}^2 + y_{mj}^2 + z_{jm}^2)^{1/2} \\
\beta_3 &= (x_{mi}^2 + y_{mi}^2 + z_{mi}^2)^{1/2} \\
\beta_4 &= (\beta_1{}^2 + \beta_2{}^2 - \beta_3{}^2)/(2\beta_1\beta_2) \\
\beta_5 &= (1.0 - \beta_4{}^2)^{1/2} \\
\beta_6 &= x_{nj}^2 + y_{nj}^2 + z_{nj}^2 \\
\beta_7 &= x_{ni}^2 + y_{ni}^2 + z_{ni}^2
\end{aligned} \tag{4.13d}$$

Note that the stiffness coefficients defined in Eq. (4.13c) apply to either a trapezium or a rectangular shear panel.

4.5 TRIANGULAR MEMBRANE ELEMENT STIFFNESS MATRIX

Figure 4.7 shows a triangular membrane element, with lumped coordinate forces and displacements positive as indicated. In practice these forces are distributed along the faces of the triangular element. Note that no generalization is lost by considering lumped forces at the nodes. This is due to the fact that the work done by these lumped forces through their respective nodal displacements is equivalent to the work done by the distributed forces through the distributed displacements of the element.

In addition, the following hypotheses are employed:

H.1. The membrane has a uniform thickness.
H.2. Bending stiffness is negligible.
H.3. The strains are constant throughout the membrane.
H.4. Linear elasticity prevails.
H.5. Material is homogeneous and isotropic.

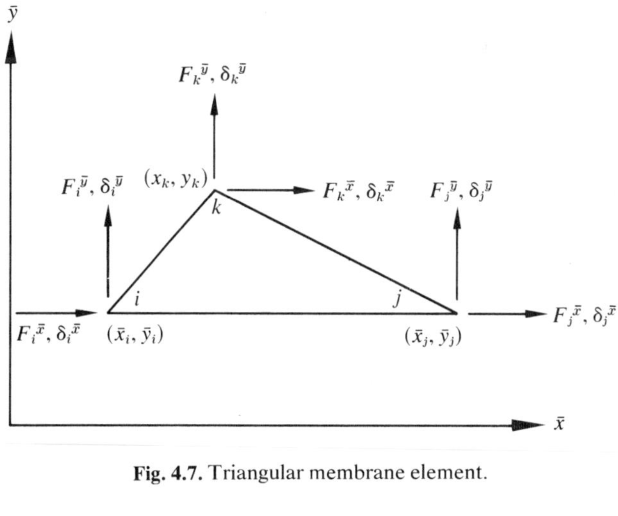

Fig. 4.7. Triangular membrane element.

For convenience, the following notations for the forces and displacements are used.

$$\begin{bmatrix} F_i^{\bar{x}} \\ F_i^{\bar{y}} \\ F_j^{\bar{x}} \\ F_j^{\bar{y}} \\ F_k^{\bar{x}} \\ F_k^{\bar{y}} \end{bmatrix}; \begin{bmatrix} \delta_i^{\bar{x}} \\ \delta_i^{\bar{y}} \\ \delta_j^{\bar{x}} \\ \delta_j^{\bar{y}} \\ \delta_k^{\bar{x}} \\ \delta_k^{\bar{y}} \end{bmatrix} = \begin{bmatrix} Q_1 \\ Q_2 \\ Q_3 \\ Q_4 \\ Q_5 \\ Q_6 \end{bmatrix}; \begin{bmatrix} q_1 \\ q_2 \\ q_3 \\ q_4 \\ q_5 \\ q_6 \end{bmatrix} \tag{4.14}$$

To derive the stiffness coefficients Eq. (3.3) will be utilized, i.e.,

$$S_{ij} = \frac{\partial^2 U}{\partial q_i \partial q_j}. \tag{3.3}$$

This means that the strain energy must be expressed in terms of the coordinate displacements.

From Eq. (2.31), for the plane strain, the strain energy in terms of strains alone is given by,

$$U = \frac{E}{2(1-\nu^2)} \iiint \left(\epsilon_{xx}^2 + \epsilon_{yy}^2 + 2\nu\epsilon_{xx}\epsilon_{yy} + \frac{1-\nu}{2}\epsilon_{xy}^2 \right) dxdydz \tag{4.15}$$

where

$$\begin{aligned} \epsilon_{xx} &= \frac{\partial q_x}{\partial x} = A_1, \\ \epsilon_{yy} &= \frac{\partial q_y}{\partial y} = B_1. \end{aligned} \tag{4.16}$$

and

$$\epsilon_{xy} = \frac{\partial q_x}{\partial y} + \frac{\partial q_y}{\partial x} = C_1.$$

where q_x and q_y are the displacements in the x and y directions, respectively.

Integrating the first two of Eq. (4.16) yields,

$$\begin{aligned} q_x &= A_1 x + F_1(y), \\ q_y &= B_1 y + F_2(x). \end{aligned} \tag{a}$$

Substituting (a) in the third of Eq. (4.16) yields,

$$\frac{\partial F_1}{\partial y} + \frac{\partial F_2}{\partial x} = C_1. \tag{b}$$

Since F_1 and F_2 are two independent functions, Eq. (b) is valid only if the following conditions hold true,

$$\frac{\partial F_2}{\partial x} = B_2, \qquad \text{and} \qquad \frac{\partial F_1}{\partial y} = A_2, \tag{c}$$

where $A_2 + B_2 = C_1 =$ constant.

Now, integrating the two parts of Eq. (c) yields,

$$\begin{aligned} F_1 &= A_2 y + A_3 \\ F_2 &= B_2 x + B_3 \end{aligned} \tag{d}$$

Substituting Eq. (d) into (a) we finally get the following displacement,

$$\begin{aligned} q_x &= A_1 x + A_2 y + A_3 \\ q_y &= B_1 y + B_2 x + B_3 \end{aligned} \tag{4.17}$$

The unknown constants in Eq. (4.17) now will be determined from the following boundary conditions (*see* Fig. 4.7),

$$\begin{array}{lll} 1.\ q_x(0,0) = q_1 & 2.\ q_x(x_2,0) = q_3 & 3.\ q_x(x_3,y_3) = q_5 \\ \quad q_y(0,0) = q_2 & \quad q_y(x_2,0) = q_4 & \quad q_y(x_3,y_3) = q_6 \end{array} \tag{4.18}$$

Substituting Eq. (4.18) into Eq. (4.17) the following set of simultaneous algebraic linear equations is obtained,

$$\begin{aligned} A_3 &= q_1 \\ B_3 &= q_2 \\ A_3 + A_1 x_2 &= q_3 \\ B_3 + B_2 x_2 &= q_4 \\ A_3 + A_2 y_3 + A_1 x_3 &= q_5 \\ B_3 + B_2 x_3 + B_1 y_3 &= q_6 \end{aligned} \tag{4.19}$$

Solving Eq. (4.19) for A_i and B_i $(i = 1, 2, 3)$ yields,

$$\begin{aligned} A_1 &= (q_3 - q_1)/x_2 \\ A_2 &= [q_5 x_2 + (x_3 - x_2) q_1 - q_3 x_3]/x_2 y_3 \\ A_3 &= q_1 \\ B_1 &= [q_6 x_2 + (x_3 - x_2) q_2 - q_4 x_3]/x_2 y_2 \\ B_2 &= (q_4 - q_2)/x_2 \\ B_3 &= q_2 \end{aligned} \tag{4.20}$$

From Eq. (4.16) the strains now become,

$$
\begin{aligned}
\epsilon_{xx} &= (q_3 - q_1)/x_2 \\
\epsilon_{yy} &= [q_6 x_2 + (x_3 - x_2) q_2 - q_4 x_3]/x_2 y_2 \\
\epsilon_{xy} &= [q_5 x_2 + (x_3 - x_2) q_1 - q_3 x_3]/x_2 y_3 + (q_4 - q_2)/x_2
\end{aligned}
\tag{4.21}
$$

Substituting Eq. (4.21) into Eq. (4.15) and integrating yield the strain energy expression in terms of nodal displacements as,

$$
\begin{aligned}
U = \frac{Etx_2 y_3}{4(1-\nu^2)} \Big\{ & (q_3 - q_1)^2/x_2^2 + [q_6 x_2 + (x_3 - x_2) q_2 - q_4 x_3]^2/x_2^2 y_3^2 \\
& + 2\nu [q_3 - q_1][q_6 x_2 + (x_3 - x_2) q_2 - q_5 x_3]/x_2^2 y_3 \\
& + \frac{1-\nu}{2} \{[q_5 x_2 + (x_3 - x_2) q_1 - q_3 x_3]/x_2 y_3 + (q_4 - q_2)/x_2\} \Big\}.
\end{aligned}
\tag{4.22}
$$

Having the expression for the strain energy in terms of the nodal coordinate displacements, the stiffness coefficients are obtained by utilizing Eq. (3.3). Hence, the stiffness matrix equation is,

$$
\begin{bmatrix} F_i^{\bar{x}} \\ F_i^{\bar{y}} \\ F_j^{\bar{x}} \\ F_j^{\bar{y}} \\ F_k^{\bar{x}} \\ F_k^{\bar{y}} \end{bmatrix}
= \bar{E}
\begin{bmatrix}
y_{ki}^2 + \nu_1 x_{kj}^2 & & & & \text{Sym.} & \\
-\nu_2 x_{kj} y_{ki} & x_{kj}^2 + \nu_1 y_{ki}^2 & & & & \\
\begin{matrix} -y_{ki}^2 \\ -\nu_1 x_{kj} x_{kj} \end{matrix} & \begin{matrix} \nu x_{kj} y_{ki} \\ +\nu_1 x_{ki} y_{ki} \end{matrix} & \begin{matrix} y_{ki}^2 \\ +\nu_1 x_{ki}^2 \end{matrix} & & & \\
\begin{matrix} \nu x_{ki} y_{ki} \\ +\nu_1 x_{kj} y_{ki} \end{matrix} & \begin{matrix} -x_{kj} x_{ki} \\ -\nu_1 y_{ki}^2 \end{matrix} & -\nu_2 x_{ki} y_{ki} & \begin{matrix} x_{ki}^2 \\ +\nu_1 y_{ki}^2 \end{matrix} & & \\
\nu_1 x_{kj} x_{ji} & -\nu_1 x_{ji} y_{ki} & -\nu_1 x_{ji} y_{ki} & \nu_1 x_{ji} y_{ki} & \nu_1 x_{ji}^2 & \\
-\nu x_{ji} y_{ki} & x_{kj} x_{ji} & \nu x_{ji} y_{ki} & x_{ji} y_{ki} & 0 & x_{ji}^2
\end{bmatrix}
\begin{bmatrix} \delta_i^{\bar{x}} \\ \delta_i^{\bar{y}} \\ \delta_j^{\bar{x}} \\ \delta_j^{\bar{y}} \\ \delta_k^{\bar{x}} \\ \delta_k^{\bar{y}} \end{bmatrix}
\tag{4.23}
$$

$$
\begin{aligned}
\text{in which } \bar{E} &= Et/2(1-\nu^2) x_{ji} y_{ki} \\
E &= \text{modulus of elasticity} \\
t &= \text{panel thickness} \\
\nu &= \text{Poisson's Ratio} \\
\nu_1 &= (1-\nu)/2 \\
\nu_2 &= (1+\nu)/2
\end{aligned}
\tag{4.23a}
$$

and

$$x_{ji} = x_j - x_i, \text{ etc.}$$

To make the stiffness matrix Eq. (4.23) applicable to elements arbitrarily oriented, the transformation technique will be employed, thus yielding,

$$\begin{bmatrix} F_i^x \\ F_i^y \\ F_i^z \\ F_j^x \\ F_j^y \\ F_j^z \\ F_k^x \\ F_k^y \\ F_k^z \end{bmatrix} = \begin{bmatrix} [T] & & & \\ & [T] & & \\ & & [T] & \\ & & & [T] \end{bmatrix}^{-1} [\bar{S}] \begin{bmatrix} [T] & & & \\ & [T] & & \\ & & [T] & \\ & & & [T] \end{bmatrix} \begin{bmatrix} \delta_i^x \\ \delta_i^y \\ \delta_i^z \\ \delta_j^x \\ \delta_j^y \\ \delta_j^z \\ \delta_k^x \\ \delta_k^y \\ \delta_k^z \end{bmatrix} \tag{4.24}$$

or in shorthand matrix notation, Eq. (4.24) becomes,

$$\{Q\} = [S]\{q\}, \tag{4.24a}$$

where

$$[S] = [T]^{-1}[\bar{S}][T]. \tag{4.24b}$$

The matrix $[\bar{S}]$ is given in Eq. (4.23) with the exception of adding rows and columns of zeros corresponding to the z-components of forces and displacements, and $[T]$ is given in either Eq. (4.9) or (4.10) depending on the sequence of rotation.

4.6 RECTANGULAR MEMBRANE ELEMENT STIFFNESS MATRIX

Figure 4.8 shows a rectangular membrane element in the local coordinate system. Lumped coordinate forces and displacements are said to be acting at all four corners and are positive as indicated. As in the

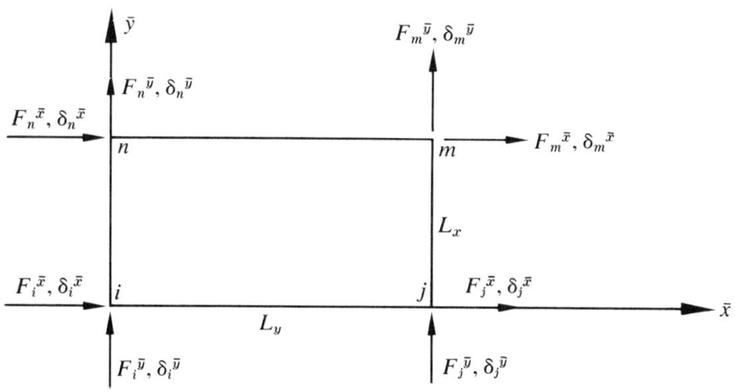

Fig. 4.8. Rectangular membrane element.

triangular element case, these forces are, in practice, distributed along the edges of the plate.

In contrast to the triangular element, the normal strains are assumed to be linear functions of x and y, while the shearing strains are constants. Hence, from Eq. (4.6) these strains can be expressed in terms of nodal displacements as follows,

$$
\begin{aligned}
\epsilon_{xx} &= (q_3 - q_1)/L_x + (q_5 - q_7 + q_1 - q_3)y/L_xL_y \\
\epsilon_{yy} &= (q_8 - q_2)/L_y + (q_6 + q_2 - q_4 - q_8)x/L_xL_y \\
\epsilon_{xy} &= (q_5 + q_7 - q_1 - q_3)/2L_y + (q_6 + q_4 - q_2 - q_8)/2L_x
\end{aligned}
\tag{4.25}
$$

where the generalized displacements q_i $(i = 1, 2, \ldots, 8)$ are defined by,

$$
\begin{bmatrix} q_1 \\ q_2 \\ q_3 \\ q_4 \\ q_5 \\ q_6 \\ q_7 \\ q_8 \end{bmatrix}
=
\begin{bmatrix} \delta_i^{\bar{x}} \\ \delta_i^{\bar{y}} \\ \delta_j^{\bar{x}} \\ \delta_j^{\bar{y}} \\ \delta_m^{\bar{x}} \\ \delta_m^{\bar{y}} \\ \delta_n^{\bar{x}} \\ \delta_n^{\bar{y}} \end{bmatrix}
$$

Substituting Eq. (4.25) into Eq. (4.15), integrating, and then utilizing Eq. (3.3) will yield the desired stiffness coefficients which are given in the following stiffness matrix equation,

$$
\begin{bmatrix} F_i^{\bar{x}} \\ F_i^{\bar{y}} \\ F_j^{\bar{x}} \\ F_j^{\bar{y}} \\ F_m^{\bar{x}} \\ F_m^{\bar{y}} \\ F_n^{\bar{x}} \\ F_n^{\bar{y}} \end{bmatrix}
= \underline{E}
\begin{bmatrix}
8+3r^2\mu_1 & & & & & & & \\
\mu_1 & 8r^2+3\mu_1 & & & & \text{Sym.} & & \\
3r^2\mu_1-8 & \mu_3 & 8+3r^2\mu_1 & & & & & \\
-3\mu_3 & 4r^2-3\mu_1 & -\mu_2 & 8r^2+3\mu_1 & & & & \\
-4-3r^2\mu_1 & -\mu_2 & 4-3r^2\mu_1 & \mu_3 & 8+3r^2\mu_1 & & & \\
-3\mu_2 & -4r^2-3\mu_1 & -\mu_3 & -8r^2+3\mu_1 & \mu_2 & 8r^2+3\mu_1 & & \\
4-3r^2\mu_1 & -\mu_3 & -4-3r^2\mu_1 & \mu_2 & -8+3m^2\mu_1 & \mu_3 & 8+3r^2\mu_1 & \\
\mu_3 & 3\mu_1-8r^2 & \mu_2 & -4r^2-3\mu_1 & -\mu_3 & 4r^2-3\mu_1 & -\mu_2 & 8r^2+3\mu_1
\end{bmatrix}
\begin{bmatrix} \delta_i^{\bar{x}} \\ \delta_i^{\bar{y}} \\ \delta_j^{\bar{x}} \\ \delta_j^{\bar{y}} \\ \delta_m^{\bar{x}} \\ \delta_m^{\bar{y}} \\ \delta_n^{\bar{x}} \\ \delta_n^{\bar{y}} \end{bmatrix}
\tag{4.26}
$$

$$\text{in which } \underline{E} = Et/24r\mu_1\mu_2$$
$$r = L_x/L_y$$
$$\mu_1 = 1-\nu \qquad (4.26a)$$
$$\mu_2 = 1+\nu$$
$$\mu_3 = 1-3\nu$$

In the reference coordinate system, Eq. (4.26) becomes,

$$\begin{bmatrix} F_i^x \\ F_i^y \\ F_i^z \\ F_j^x \\ F_j^y \\ F_j^z \\ F_m^x \\ F_m^y \\ F_m^z \\ F_n^x \\ F_n^y \\ F_n^z \end{bmatrix} = \begin{bmatrix} [T] & & & \\ & [T] & & \\ & & [T] & \\ & & & [T] \end{bmatrix}^{-1} [\bar{S}] \begin{bmatrix} [T] & & & \\ & [T] & & \\ & & [T] & \\ & & & [T] \end{bmatrix} \begin{bmatrix} \delta_i^x \\ \delta_i^y \\ \delta_i^z \\ \delta_j^x \\ \delta_j^y \\ \delta_j^z \\ \delta_m^x \\ \delta_m^y \\ \delta_m^z \\ \delta_n^x \\ \delta_n^y \\ \delta_n^z \end{bmatrix} \qquad (4.27)$$

where $[T]$ is given in either Eq. (4.9) or (4.10) depending on the element sequence rotations, and $[\bar{S}]$ is given in Eq. (4.26) with the exception of adding rows and columns of zeros corresponding to the force and displacement components in the z-direction.

Illustrative Examples

Example 1

Find the displacements and internal loads for the structure shown below. The cross-sectional area of each rod is assumed to be 1.0 in.^{2} and $E = 10.3 \times 10^6$ psi, the thickness of the shear panel is 0.04 in. and $G = 3 \times 10^6$ psi.

The stiffness matrix equations for rod elements 1, 2 and 3 are from Eq. (3.17):

Rod 1

$$\begin{bmatrix} F_1^x \\ F_1^y \\ F_3^x \\ F_3^y \end{bmatrix} = 10^5 \begin{bmatrix} 0 & 0 & 0 & 0 \\ 0 & 6.8 & 0 & -6.8 \\ 0 & 0 & 0 & 0 \\ 0 & -6.8 & 0 & 6.8 \end{bmatrix} \begin{bmatrix} \delta_1^x \\ \delta_1^y \\ \delta_3^x \\ \delta_3^y \end{bmatrix} \qquad (a)$$

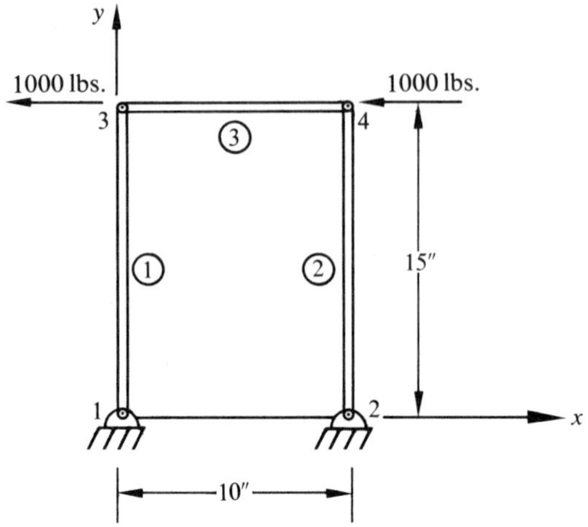

Rod 2

$$\begin{bmatrix} F_2^x \\ F_2^y \\ F_4^x \\ F_4^y \end{bmatrix} = 10^5 \begin{bmatrix} 0 & 0 & 0 & 0 \\ 0 & 6.8 & 0 & -6.8 \\ 0 & 0 & 0 & 0 \\ 0 & -6.8 & 0 & 6.8 \end{bmatrix} \begin{bmatrix} \delta_2^x \\ \delta_2^y \\ \delta_4^x \\ \delta_4^y \end{bmatrix} \tag{b}$$

Rod 3

$$\begin{bmatrix} F_3^x \\ F_3^y \\ F_4^x \\ F_4^y \end{bmatrix} = 10^5 \begin{bmatrix} 10.3 & 0 & -10.3 & 0 \\ 0 & 0 & 0 & 0 \\ -10.3 & 0 & 10.3 & 0 \\ 0 & 0 & 0 & 0 \end{bmatrix} \begin{bmatrix} \delta_3^x \\ \delta_3^y \\ \delta_4^x \\ \delta_4^y \end{bmatrix} \tag{c}$$

From Eq. (4.7) the shear panel stiffness matrix equation is,

$$\begin{bmatrix} F_1^x \\ F_1^y \\ F_2^x \\ F_2^y \\ F_4^x \\ F_4^y \\ F_3^x \\ F_3^y \end{bmatrix} = 10^5 \begin{bmatrix} 0.2 & 0.3 & 0.2 & -0.3 & -0.2 & -0.3 & -0.2 & 0.3 \\ 0.3 & 0.45 & 0.3 & -0.45 & -0.3 & -0.45 & -0.3 & -0.45 \\ 0.2 & 0.3 & 0.2 & -0.3 & -0.2 & -0.3 & -0.2 & 0.3 \\ -0.3 & -0.45 & -0.3 & -0.45 & 0.3 & 0.45 & 0.3 & -0.45 \\ -0.2 & -0.3 & -0.2 & -0.3 & 0.2 & 0.3 & 0.2 & -0.3 \\ -0.3 & -0.45 & -0.3 & 0.45 & 0.3 & 0.45 & 0.3 & -0.45 \\ -0.2 & -0.3 & -0.2 & 0.3 & 0.2 & 0.3 & 0.2 & -0.3 \\ 0.3 & 0.45 & 0.3 & -0.45 & -0.3 & -0.45 & -0.3 & 0.45 \end{bmatrix} \begin{bmatrix} \delta_1^x \\ \delta_1^y \\ \delta_2^x \\ \delta_2^y \\ \delta_4^x \\ \delta_4^y \\ \delta_3^x \\ \delta_3^y \end{bmatrix} \tag{d}$$

Superposing Eqs. (a), (b), (c), and (d) to get the matrix equation for the structure, introducing the boundary conditions ($\delta_1^x = \delta_1^y = \delta_2^x = \delta_2^y = 0$) and deleting rows and columns corresponding to the zero displacements yield the following reduced aggregate stiffness matrix equation,

$$\begin{bmatrix} -1000 \\ 0 \\ -1000 \\ 0 \end{bmatrix} = 10^5 \begin{bmatrix} 10.5 & -0.3 & -10.1 & 0.3 \\ -0.3 & 7.32 & -0.3 & -0.45 \\ -10.1 & -0.3 & 10.5 & 0.3 \\ 0.3 & -0.45 & 0.3 & 7.32 \end{bmatrix} \begin{bmatrix} \delta_4^x \\ \delta_4^y \\ \delta_3^x \\ \delta_3^y \end{bmatrix} \tag{e}$$

Upon inverting the above matrix the following is obtained.

$$\begin{bmatrix} \delta_4^x \\ \delta_4^y \\ \delta_3^x \\ \delta_3^y \end{bmatrix} = 10^{-5} \begin{bmatrix} 1.438 & 0.109 & 1.39 & -0.109 \\ 0.109 & 0.146 & 0.109 & 0 \\ 1.39 & 0.109 & 1.438 & -0.109 \\ -0.109 & 0 & -0.109 & 0.146 \end{bmatrix} \begin{bmatrix} -1000 \\ 0 \\ -1000 \\ 0 \end{bmatrix} \tag{f}$$

Solving for the displacements we get,

$$\begin{aligned} \delta_3^x &= -2.828 \times 10^{-2} \text{ in.} \qquad & \delta_4^x &= -2.828 \times 10^{-2} \text{ in.} \\ \delta_3^y &= -2.184 \times 10^{-3} \text{ in.} \qquad & \delta_4^y &= -2.184 \times 10^{-3} \text{ in.} \end{aligned} \tag{g}$$

Upon substitution, Eq. (g) into the individual element matrix equations (a–d) and solving for the internal loads yield,

Rod 1 $\quad F_1^x = F_3^x = 0, \quad F_1^y = 1500 \text{ lb}, \quad F_3^y = -1500 \text{ lb}$

Rod 2 $\quad F_2^x = F_4^x = 0, \quad F_2^y = -1500 \text{ lb}, \quad F_4^y = 1500 \text{ lb}$

Rod 3 $\quad F_3^y = F_4^x = 0, \quad F_3^x = F_4^x, \quad 0$

Panel:

$$\begin{aligned} F_1^x &= 1000 \text{ lb}, \qquad & F_1^y &= 1500 \text{ lb} \\ F_2^x &= 1000 \text{ lb}, \qquad & F_2^y &= -1500 \text{ lb} \\ F_3^x &= -1000 \text{ lb}, \qquad & F_3^y &= 1500 \text{ lb} \\ F_4^x &= -1000 \text{ lb}, \qquad & F_4^y &= -1500 \text{ lb} \end{aligned}$$

The shear flow is shown on the sketch below.

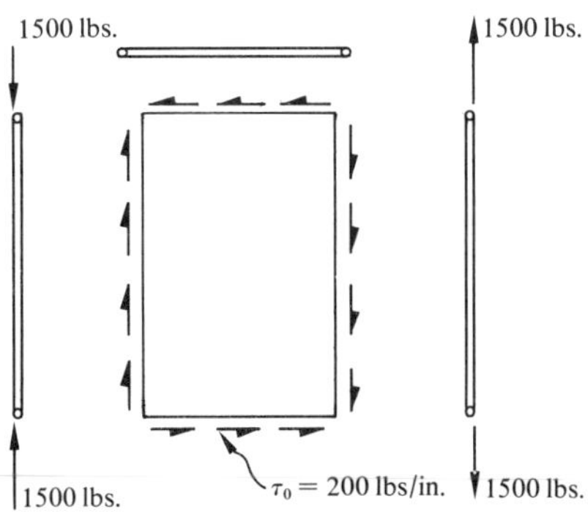

Example 2

Shown below is a two-bay beam. It is required to determine the displacements and the shear flow in the webs. Assume that each stiffener (rod) has a cross-sectional area of 0.2 in².

$$E = 10 \times 10^6 \text{ psi}$$
$$G = 4 \times 10^6 \text{ psi}$$

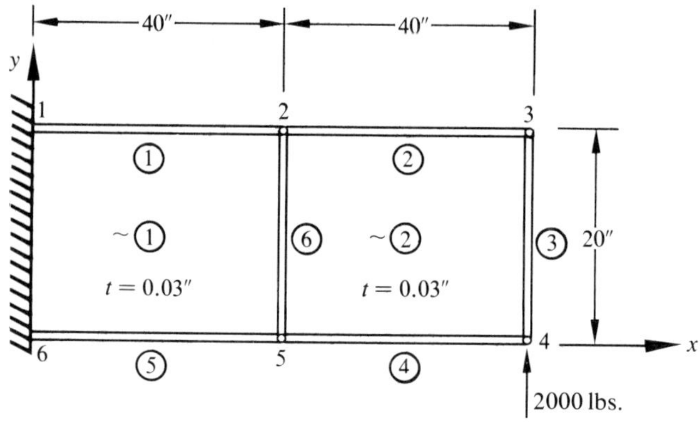

Following the same procedure as presented in Example 1, the following reduced aggregate stiffness matrix equation for the structure is obtained,

$$\begin{bmatrix} 0 \\ 0 \\ 0 \\ 0 \\ 0 \\ 2000 \\ 0 \\ 0 \end{bmatrix} = 10^4 \begin{bmatrix} 22 & & & & & & & \\ 0 & 13 & & & & \text{Sym.} & & \\ 1 & -3 & 11 & & & & & \\ 3 & -1.5 & 3 & 11.5 & & & & \\ -6 & 3 & -6 & -3 & 11 & & & \\ 3 & -1.5 & 3 & -8.5 & -3 & 11.5 & & \\ -12 & 0 & -6 & -3 & 1 & -3 & 22 & \\ 0 & -7 & -3 & -1.5 & 3 & -1.5 & 0 & 13 \end{bmatrix} \begin{bmatrix} \delta_2^x \\ \delta_2^y \\ \delta_3^x \\ \delta_3^y \\ \delta_4^x \\ \delta_4^y \\ \delta_5^x \\ \delta_5^y \end{bmatrix} \tag{a}$$

Upon inversion Eq. (a) becomes,

$$\begin{bmatrix} \delta_2^x \\ \delta_2^y \\ \delta_3^x \\ \delta_3^y \\ \delta_4^x \\ \delta_4^y \\ \delta_5^x \\ \delta_5^y \end{bmatrix} = 10^{-4} \begin{bmatrix} 0.2 & & & & & & & \\ -0.2 & 0.6 & & & & \text{Sym.} & & \\ 0.2 & -0.2 & 0.4 & & & & & \\ -0.6 & 1.37 & -0.8 & 4.36 & & & & \\ 0 & 0.2 & 0 & 0.8 & 0.4 & & & \\ -0.6 & 1.37 & -0.8 & 4.31 & 0.8 & 4.36 & & \\ 0 & 0.2 & 0 & 0.6 & 0.2 & 0.6 & 0.2 & \\ -0.2 & 0.54 & -0.2 & 1.37 & 0.2 & 1.37 & 0.2 & 0.6 \end{bmatrix} \begin{bmatrix} 0 \\ 0 \\ 0 \\ 0 \\ 0 \\ 2000 \\ 0 \\ 0 \end{bmatrix} \tag{b}$$

Solving for the displacements yields,

$$\delta_2{}^x = -0.12 \text{ in.}, \qquad \delta_4{}^x = 0.16 \text{ in.}$$
$$\delta_2{}^y = 0.274 \text{ in.}, \qquad \delta_4{}^y = 0.872 \text{ in.}$$
$$\delta_3{}^x = -0.16 \text{ in.}, \qquad \delta_5{}^x = 0.12 \text{ in.}$$
$$\delta_3{}^y = 0.862 \text{ in.}, \qquad \delta_5{}^y = 0.274 \text{ in.}$$

The calculated internal loads are shown on the sketch below.

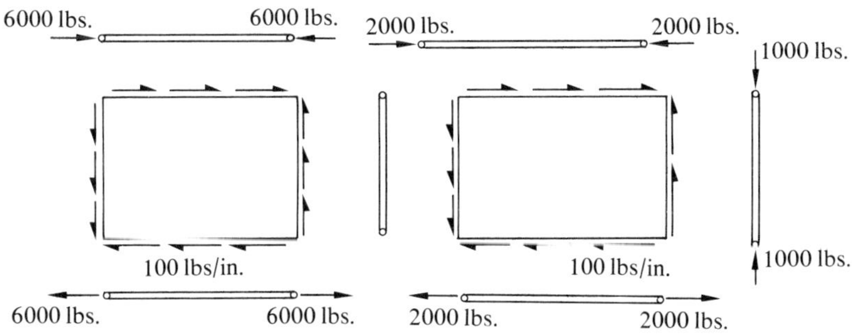

Example 3

Shown in the figure below is a structure which consists of two trapezium shear panels stiffened by axial rods. Assume that the cross-sectional area of each rod is 1.0 in.^{2} and the thickness of each panel is 0.04 in. Determine the displacements and nodal internal loads. $E = 10.3 \times 10^6$ psi. $G = 3 \times 10^6$ psi.

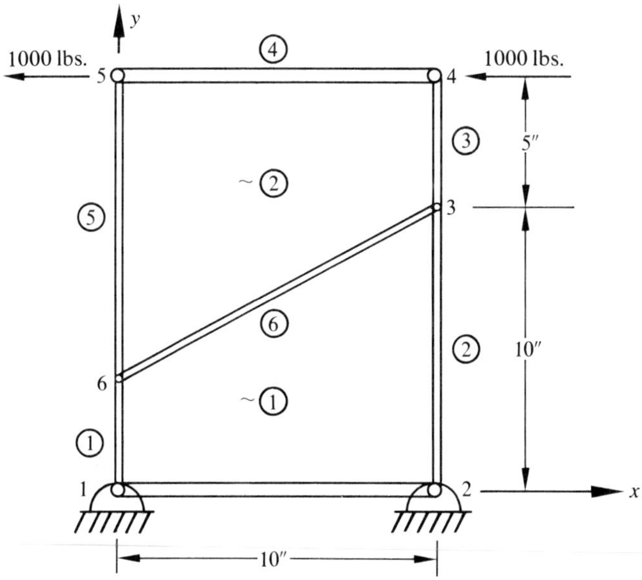

The element matrix equations are,

Rod 1

$$\begin{bmatrix} F_1^x \\ F_1^y \\ F_6^x \\ F_6^y \end{bmatrix} = 10^6 \begin{bmatrix} 0 & 0 & 0 & 0 \\ 0 & 2.06 & 0 & -2.06 \\ 0 & 0 & 0 & 0 \\ 0 & -2.06 & 0 & 2.06 \end{bmatrix} \begin{bmatrix} \delta_1^x \\ \delta_1^y \\ \delta_6^x \\ \delta_6^y \end{bmatrix} \tag{a}$$

Rod 2

$$\begin{bmatrix} F_2^x \\ F_2^y \\ F_3^x \\ F_3^y \end{bmatrix} = 10^6 \begin{bmatrix} 0 & 0 & 0 & 0 \\ 0 & 1.03 & 0 & -1.03 \\ 0 & 0 & 0 & 0 \\ 0 & -1.03 & 0 & 1.03 \end{bmatrix} \begin{bmatrix} \delta_2^x \\ \delta_2^y \\ \delta_3^x \\ \delta_3^y \end{bmatrix} \tag{b}$$

Rod 3

$$\begin{bmatrix} F_3^x \\ F_3^y \\ F_4^x \\ F_4^y \end{bmatrix} = 10^6 \begin{bmatrix} 0 & 0 & 0 & 0 \\ 0 & 2.06 & 0 & -2.06 \\ 0 & 0 & 0 & 0 \\ 0 & -2.06 & 0 & 2.06 \end{bmatrix} \begin{bmatrix} \delta_3^x \\ \delta_3^y \\ \delta_4^x \\ \delta_4^y \end{bmatrix} \tag{c}$$

Rod 4

$$\begin{bmatrix} F_4^x \\ F_4^y \\ F_5^x \\ F_5^y \end{bmatrix} = 10^6 \begin{bmatrix} 1.03 & 0 & -1.03 & 0 \\ 0 & 0 & 0 & 0 \\ -1.03 & 0 & 1.03 & 0 \\ 0 & 0 & 0 & 0 \end{bmatrix} \begin{bmatrix} \delta_4^x \\ \delta_4^y \\ \delta_5^x \\ \delta_5^y \end{bmatrix} \tag{d}$$

Rod 5

$$\begin{bmatrix} F_5^x \\ F_5^y \\ F_6^x \\ F_6^y \end{bmatrix} = 10^6 \begin{bmatrix} 0 & 0 & 0 & 0 \\ 0 & 1.03 & 0 & -1.03 \\ 0 & 0 & 0 & 0 \\ 0 & -1.03 & 0 & 1.03 \end{bmatrix} \begin{bmatrix} \delta_5^x \\ \delta_5^y \\ \delta_6^x \\ \delta_6^y \end{bmatrix} \tag{e}$$

Rod 6

$$\begin{bmatrix} F_3^x \\ F_3^y \\ F_6^x \\ F_6^y \end{bmatrix} = 10^6 \begin{bmatrix} 0.74 & 0.37 & -0.74 & -0.37 \\ 0.37 & 0.18 & -0.37 & -0.18 \\ -0.74 & -0.37 & 0.74 & 0.37 \\ -0.37 & -0.18 & 0.37 & 0.18 \end{bmatrix} \begin{bmatrix} \delta_3^x \\ \delta_3^y \\ \delta_6^x \\ \delta_6^y \end{bmatrix} \tag{f}$$

Panel 1

$$\begin{bmatrix} F_1^x \\ F_1^y \\ F_2^x \\ F_2^y \\ F_3^x \\ F_3^y \\ F_6^x \\ F_6^y \end{bmatrix} = 10^6 \begin{bmatrix} 0.04 & & & & & & & \\ 0.02 & 0.01 & & & \text{Sym.} & & & \\ 0.04 & 0.02 & 0.04 & & & & & \\ -0.04 & -0.02 & -0.04 & 0.04 & & & & \\ -0.04 & -0.02 & -0.04 & 0.04 & 0.04 & & & \\ -0.02 & -0.01 & -0.02 & 0.02 & 0.02 & 0.04 & & \\ -0.04 & -0.01 & -0.04 & 0.04 & 0.04 & 0.02 & 0.04 & \\ 0.04 & 0.02 & 0.04 & -0.04 & -0.04 & -0.02 & -0.04 & 0.04 \end{bmatrix} \begin{bmatrix} \delta_1^x \\ \delta_1^y \\ \delta_2^x \\ \delta_2^y \\ \delta_3^x \\ \delta_3^y \\ \delta_6^x \\ \delta_6^y \end{bmatrix} \tag{g}$$

Panel 2

$$\begin{bmatrix} F_6^x \\ F_6^y \\ F_3^x \\ F_3^y \\ F_4^x \\ F_4^y \\ F_5^x \\ F_5^y \end{bmatrix} = 10^6 \begin{bmatrix} 0.33 & & & & & & & \\ -0.17 & 0.087 & & & \text{Sym.} & & & \\ -0.09 & 0.047 & 0.025 & & & & & \\ 0.013 & 0.007 & 0.004 & 0.0005 & & & & \\ -0.33 & 0.17 & 0.09 & 0.013 & 0.33 & & & \\ 0.17 & -0.087 & -0.047 & -0.007 & -0.17 & 0.087 & & \\ 0.09 & -0.047 & -0.025 & -0.004 & -0.09 & 0.047 & 0.025 & \\ 0.013 & -0.007 & -0.004 & -0.0005 & -0.013 & 0.007 & 0.004 & 0.0005 \end{bmatrix} \begin{bmatrix} \delta_6^x \\ \delta_6^y \\ \delta_3^x \\ \delta_3^y \\ \delta_4^x \\ \delta_4^y \\ \delta_5^x \\ \delta_5^y \end{bmatrix} \tag{h}$$

Superposing Eqs. (a–h), introducing the zero boundary conditions ($\delta_1^x = \delta_1^y = \delta_2^x = \delta_2^y = 0$), and deleting the rows and columns corresponding to zero displacements will yield,

$$\begin{bmatrix} 0 \\ 0 \\ -1000 \\ 0 \\ -1000 \\ 0 \\ 0 \\ 0 \end{bmatrix} = 10^6 \begin{bmatrix} 0.802 & & & & & & & \\ 0.392 & 3.285 & & & \text{Sym.} & & & \\ 0.091 & 0.013 & 1.359 & & & & & \\ -0.047 & -2.067 & -0.169 & 2.147 & & & & \\ -0.025 & -0.004 & -1.121 & 0.047 & 1.055 & & & \\ -0.004 & -0.0005 & -0.013 & 0.007 & 0.004 & 1.03 & & \\ -0.788 & -0.362 & -0.329 & 0.169 & 0.091 & 0.013 & 1.11 & \\ -0.362 & -0.198 & 0.169 & -0.087 & -0.047 & -1.04 & 0.16 & 3.401 \end{bmatrix} \begin{bmatrix} \delta_3^x \\ \delta_3^y \\ \delta_4^x \\ \delta_4^y \\ \delta_5^x \\ \delta_5^y \\ \delta_6^x \\ \delta_6^y \end{bmatrix} \tag{i}$$

Upon performing the inversion in the above matrix, the following is obtained,

$$\begin{bmatrix} \delta_3^x \\ \delta_3^y \\ \delta_4^x \\ \delta_4^y \\ \delta_5^x \\ \delta_5^y \\ \delta_6^x \\ \delta_6^y \end{bmatrix} = 10^{-6} \begin{bmatrix} 7.1 & & & & & & & \\ -0.49 & 0.97 & & & \text{Sym.} & & & \\ 5.02 & 0.82 & 14.4 & & & & & \\ -0.49 & 0.97 & 1.17 & 1.46 & & & & \\ 5.02 & 0.82 & 14.8 & 1.17 & 16.1 & & & \\ 0.37 & 0 & -0.24 & 0 & -0.24 & 1.46 & & \\ 6 & 0 & 6.78 & 0 & 6.78 & 0.12 & 6.62 & \\ 0.37 & 0 & -0.29 & 0 & -0.29 & 0.47 & 0.12 & 0.47 \end{bmatrix} \begin{bmatrix} 0 \\ 0 \\ -1000 \\ 0 \\ -1000 \\ 0 \\ 0 \\ 0 \end{bmatrix} \tag{j}$$

Solving for the displacements.

$$\delta_3^x = -1.004 \times 10^{-2} \text{ in.}, \qquad \delta_4^x = -2.916 \times 10^{-2} \text{ in.}$$

$$\delta_3^y = -1.649 \times 10^{-3} \text{ in.}, \qquad \delta_4^y = -2.34 \times 10^{-3} \text{ in.}$$

$$\delta_5^x = -3.087 \times 10^{-2} \text{ in.}, \qquad \delta_6^x = -1.355 \times 10^{-2} \text{ in.}$$

$$\delta_5^y = 4.753 \times 10^{-4} \text{ in.}, \qquad \delta_6^y = 5.82 \times 10^{-4} \text{ in.}$$

Having determined the displacements, the internal nodal forces may be obtained from the individual element stiffness matrix equations. The results are shown on the sketch below.

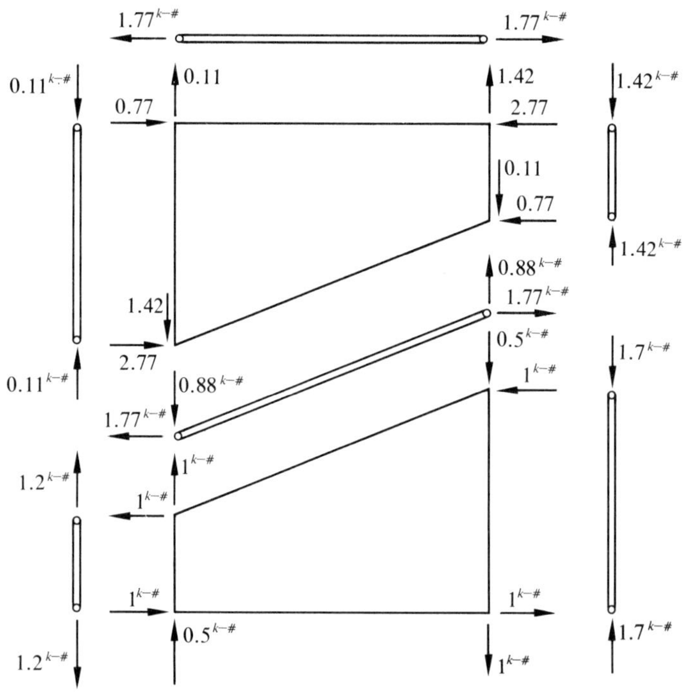

Example 4

Determine the displacements on the box-beam shown below. Assume that each stiffner has a cross-sectional area of 0.5 in^2 and the thickness of each panel is 0.032 in. $E = 10.3 \times 10^6$ psi. $G = 3 \times 10^6$ psi.

In a similar manner to the previous examples, the reduced aggregate stiffness matrix equation may be obtained and is given below.

$$
\begin{bmatrix} 0 \\ 0 \\ 1000 \\ 0 \\ 0 \\ -1000 \\ 0 \\ 0 \\ 0 \\ 0 \\ 0 \\ 0 \end{bmatrix}
= 10^5
\begin{bmatrix}
5.97 & & & & & & & & & & & \\
-0.24 & 7.86 & & & & & & \text{Sym.} & & & & \\
-0.24 & 0.24 & 10.6 & & & & & & & & & \\
-0.34 & 0.24 & 0 & 5.97 & & & & & & & & \\
-0.24 & -6.85 & 0.24 & 0.24 & 7.86 & & & & & & & \\
0 & -0.24 & -0.17 & -0.24 & -0.24 & 10.6 & & & & & & \\
-0.48 & 0 & 0.24 & 0 & 0 & 0 & 5.97 & & & & & \\
0 & -0.34 & -0.24 & 0 & -0.34 & 0.24 & -0.24 & 7.86 & & & & \\
-0.24 & 0.24 & -10 & 0 & 0.24 & -0.17 & 0.24 & -0.24 & 10.6 & & & \\
0 & 0 & 0 & -0.48 & 0 & 0.24 & -0.35 & 0.24 & 0 & 5.97 & & \\
0 & -0.34 & -0.24 & 0 & -0.34 & 0.24 & -0.24 & -6.85 & -0.24 & 0.24 & 7.86 & \\
0 & -0.24 & -0.17 & -0.24 & -0.24 & -10 & 0 & 0.24 & -0.17 & 0.24 & 0.24 & 10.6
\end{bmatrix}
\begin{bmatrix} \delta_2^x \\ \delta_2^y \\ \delta_2^z \\ \delta_3^x \\ \delta_3^y \\ \delta_3^z \\ \delta_6^x \\ \delta_6^y \\ \delta_6^z \\ \delta_7^x \\ \delta_7^y \\ \delta_7^z \end{bmatrix}
\quad \text{(a)}
$$

Upon inversion. Eq. (a) becomes.

$$
\begin{bmatrix} \delta_2^x \\ \delta_2^y \\ \delta_2^z \\ \delta_3^x \\ \delta_3^y \\ \delta_3^z \\ \delta_6^x \\ \delta_6^y \\ \delta_6^z \\ \delta_7^x \\ \delta_7^y \\ \delta_7^z \end{bmatrix}
= 10^{-6}
\begin{bmatrix}
1.84 & & & & & & & & & & & \\
0.68 & 12.42 & & & & & & & & & & \\
1.18 & -3.79 & 19.34 & & & & & & \text{Sym.} & & & \\
0.10 & -0.68 & 0.76 & 1.84 & & & & & & & & \\
0.68 & 11.74 & -3.79 & -0.68 & 12.42 & & & & & & & \\
0.76 & 3.79 & 5.62 & 1.18 & 3.79 & 19.34 & & & & & & \\
0.10 & 0.70 & -1.18 & -0.10 & 0.70 & -0.76 & 1.84 & & & & & \\
0.70 & 4.78 & 3.79 & -0.70 & 4.78 & -3.79 & 0.68 & 12.42 & & & & \\
1.18 & -3.79 & 18.86 & 0.76 & -3.79 & 5.62 & -1.18 & 3.79 & 19.34 & & & \\
-0.10 & -0.70 & -0.76 & 0.10 & -0.70 & -1.18 & 0.10 & -0.68 & -0.76 & 1.84 & & \\
0.70 & 4.78 & 3.79 & -0.70 & 4.78 & -3.79 & 0.68 & 11.74 & 3.79 & -0.68 & 12.42 & \\
0.76 & 3.79 & 5.62 & 1.18 & 3.79 & 18.86 & -0.76 & -3.79 & 5.62 & -1.18 & -0.79 & 19.34
\end{bmatrix}
\begin{bmatrix} 0 \\ 0 \\ 1000 \\ 0 \\ 0 \\ -1000 \\ 0 \\ 0 \\ 0 \\ 0 \\ 0 \\ 0 \end{bmatrix}
\quad \text{(b)}
$$

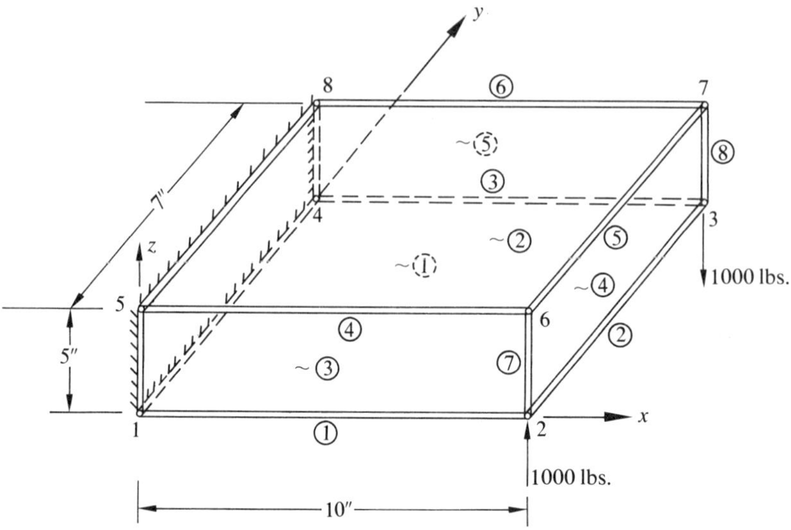

Solving for the displacements Eq. (b) yields,

$\delta_2^x = 4.16 \times 10^{-4}$ in., $\quad \delta_3^x = -4.16 \times 10^{-4}$ in.
$\delta_2^y = 7.58 \times 10^{-3}$ in., $\quad \delta_3^y = -7.58 \times 10^{-3}$ in.
$\delta_2^z = 1.37 \times 10^{-2}$ in., $\quad \delta_3^z = -1.37 \times 10^{-2}$ in.
$\delta_6^x = -4.16 \times 10^{-4}$ in., $\quad \delta_7^x = 4.16 \times 10^{-4}$ in.
$\delta_6^y = 7.58 \times 10^{-3}$ in., $\quad \delta_7^y = 7.58 \times 10^{-3}$ in.
$\delta_6^z = 1.32 \times 10^{-2}$ in., $\quad \delta_7^z = -1.32 \times 10^{-2}$ in.

4.7 SUPPLEMENTARY PROBLEMS

Problem 4.1. Determine the internal forces on the beam shown by the stiffness matrix method. Assume that shear is taken by the panel and normal stresses are taken by the rods. The panel thickness is 0.01 in. and each rod has an area of 1 in^2. $E = 10^7$ psi, $G = 4 \times 10^6$ psi.

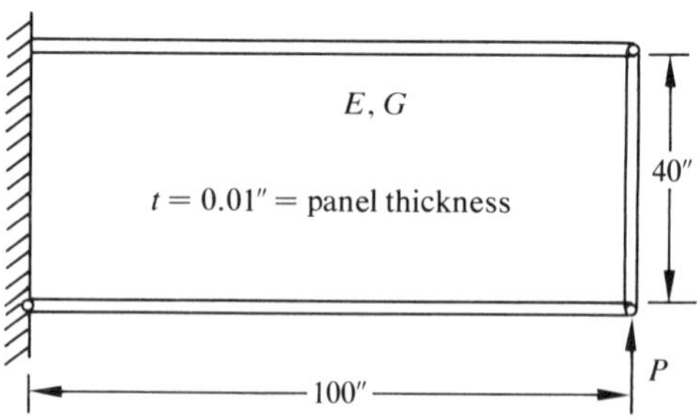

Problem 4.2. Generate the reduced aggregate stiffness matrix for the beam shown below. A = cross-sectional area of stringers, t = thickness of panel.

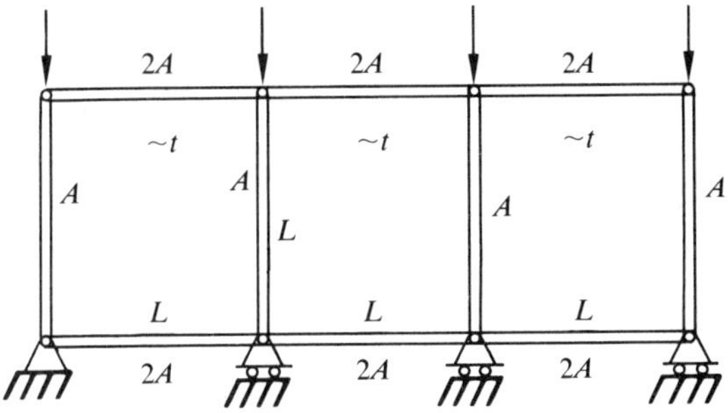

panel thickness $= t =$ constant.

Problem 4.3. Generate the aggregate stiffness matrix for the box beam shown. Assume constant thickness for all skins, and constant areas for the stringers.

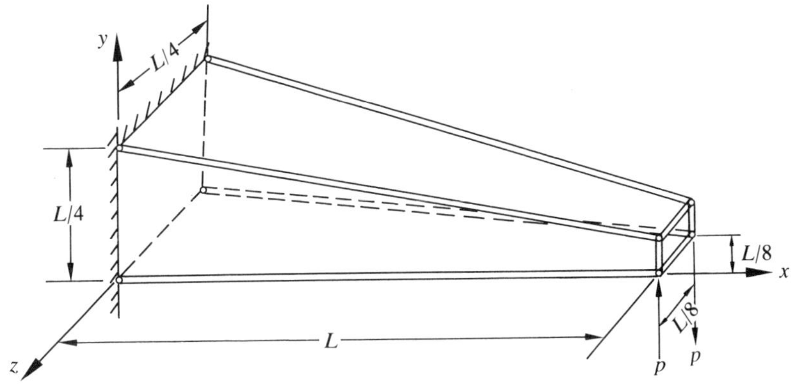

Problem 4.4. Using the stiffness method, determine the internal forces on the structure given below. $t = 0.04$ in., $G = 4 \times 10^5$ psi, $E = 3 \times 10^6$ psi, $L = 90$ in., $A = 0.2$ in^2.

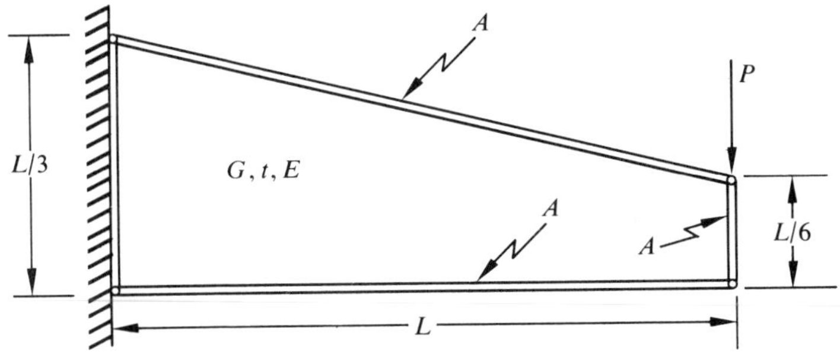

Problem 4.5. Derive the transformation matrix of a rectangular element in terms of its nodal coordinates. Check results against Eqs. (4.7a, b, c, and 4.8a).

Problem 4.6. Derive the stiffness matrix of a triangular shear panel.

Problem 4.7. Given a plate structure under the action of in-plane tensile forces only, solve for the internal stresses: (a) by idealizing the plate into four rectangular membranes; (b) by idealizing the plate into four triangular membranes.

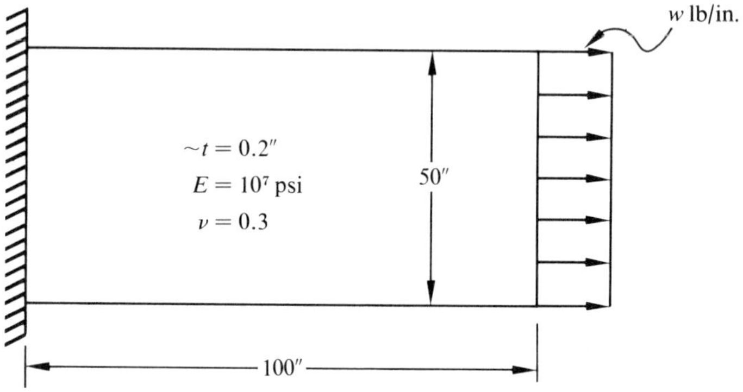

Problem 4.8. Given a plate structure under the action of in-plane loads only, determine the reduced aggregate stiffness matrix equation.

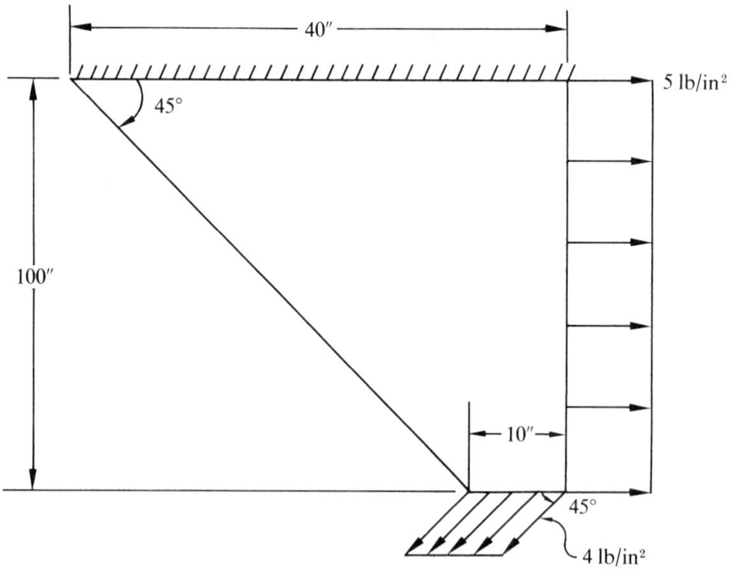

Problem 4.9. Derive the stiffness matrix equation of a triangular membrane element if six nodes are considered at the boundary instead of three nodes.

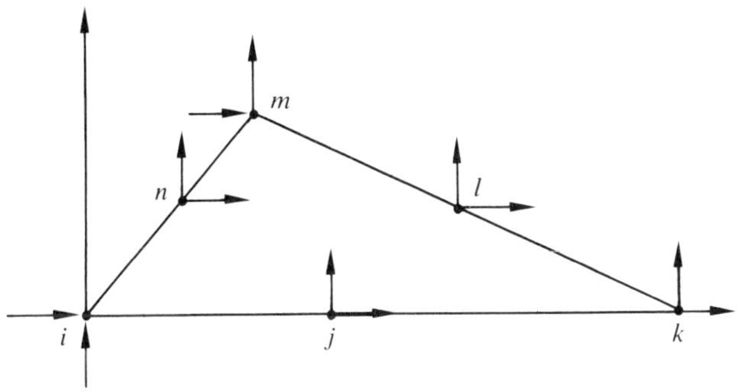

Problem 4.10. Determine all the displacements and the internal loads for the box-beam shown. Assume that each stiffener has a cross-sectional area of 0.1 in^2 and each shear panel has a thickness of 0.02 in. $E = 10^7$ psi, $G = 4 \times 10^6$ psi.

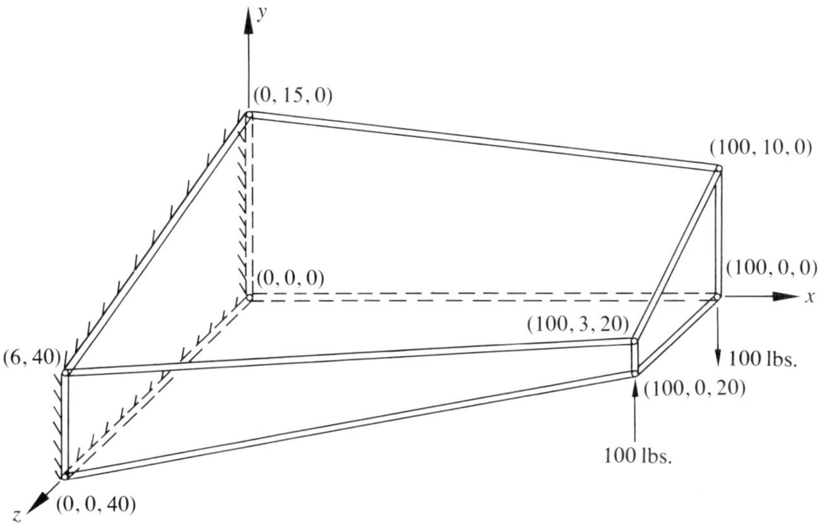

Problem 4.11. Derive the stiffness matrix equation for a rectangular membrane element if eight nodes are considered on the boundary instead of four nodes as previously dealt with.

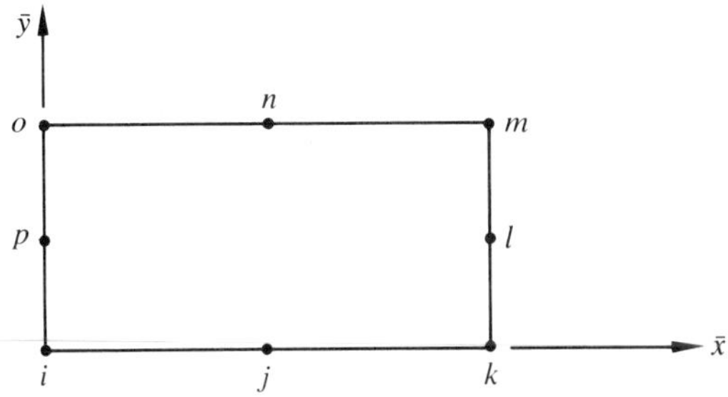

CHAPTER FIVE

Matrix Force Method

5.1 INTRODUCTION

The force method of structural analysis is based on the equation of equilibrium and compatability conditions of displacements. In this chapter the theory and the application of this method are presented.

In general, structures can be classified into two categories:

1. Statically determinate structures.
2. Statically indeterminate or redundant structures.

For the second, the equations of equilibrium are insufficient in determining internal and reactive forces on the structural system. Thus, the compatability conditions of displacements must be utilized to furnish additional equations before a solution can be obtained.

5.2 GENERAL THEORY OF THE FORCE METHOD

As in the stiffness matrix method, the actual structure must be idealized into a mathematical model which consists of distinct structural elements interconnected through node points.

If we represent the external applied nodal forces on the structure by the force vector, $\{Q\}$,

$$\{Q\} = \lfloor Q_1 Q_2 Q_3, \ldots, Q_i \rfloor^T, \tag{5.1}$$

then the induced elastic displacements corresponding to and in the direction of these forces can be written as,

$$\{q\} = \lfloor q_1 q_2 q_3, \ldots, q_i \rfloor^T. \tag{5.2}$$

The superscript, T, here indicates transpose.

Let us assume that the structural system consists of m redundants, this is equivalent to stating that there are m redundants which are unknown forces to be determined, i.e.,

$$\{R\} = [R_1 R_2 R_3, \ldots, R_m]^T, \tag{5.3}$$

which may be redundant restraint forces (reactions) or redundant element internal forces, or a combination of both. If such redundants are removed (by either cutting out certain elements or removing certain restraints or both) then, the system becomes a statically determinate one. The force components of the redundant force vector, $\{R\}$, are in reality unknowns and will be treated as unknown external applied forces. These must be determined if a solution of a structural problem is to be achieved.

The magnitude and directions of these redundant forces can be determined from the compatability conditions, i.e., the relative displacements at each "cut" must vanish.

It should be observed at this point that the number of subscripts on $\{R\}$, $\{q\}$, and $\{Q\}$ has no relation to the structure node points. These subscripts merely represent the number of external applied loads in the case of $\{Q\}$, and the number of member internal redundant forces in the case of $\{R\}$.

If the actual internal member forces are represented by the vector $\{I\}$ where

$$\{I\} = \lfloor I_1 I_2 I_3, \ldots, I_j \rfloor^T, \tag{5.4}$$

then, for linearly elastic structures, $\{I\}$ can be related to $\{Q\}$ and $\{R\}$ as follows,

$$\begin{bmatrix} I_1 \\ I_2 \\ I_3 \\ \vdots \\ I_j \end{bmatrix} = [B_0] \begin{bmatrix} Q_1 \\ Q_2 \\ Q_3 \\ \vdots \\ Q_j \end{bmatrix} + [B_1] \begin{bmatrix} R_1 \\ R_2 \\ R_3 \\ \vdots \\ R_j \end{bmatrix} \tag{5.5}$$

In matrix compact notation, Eq. (5.5) becomes,

$$\{I\} = [B_0]\{Q\} + [B_1]\{R\}, \tag{5.5a}$$

or

$$\{I\} = \Big[[B_0] \Big| [B_1] \Big] \left\{ \frac{Q}{R} \right\}, \tag{5.5b}$$

where, in general, $[B_0]$ and $[B_1]$ are rectangular matrices where coefficients are obtained from the equilibrium conditions on the structure. For instance, if Q_i is set to be equal to a unit load with all other loads including R_m held at zero, the resultant internal forces represent the coefficients corresponding to the ith column in the $[B_0]$ matrix. Likewise, the internal forces which result from a unit load R_m with all others held at zero represent the coefficients corresponding to the mth column in the $[B_1]$ matrix. Usually, Eq. (5.5) is referred to as the general matrix form of the equilibrium equations.

To formulate the compatibility conditions the complementary strain energy will be utilized. In terms of element flexibilities this energy may be expressed as,

$$U = \tfrac{1}{2}\lfloor I_1, I_2, \ldots, I_j \rfloor \begin{bmatrix} [F_1] & & & & \\ & [F_2] & & & \\ & & \cdot & & \\ & & & \cdot & \\ & & & & [F_j] \end{bmatrix} \begin{bmatrix} I_1 \\ I_2 \\ \vdots \\ I_j \end{bmatrix} \tag{5.6}$$

where $[I]$ and $[F]$ represent the element internal forces and flexibility matrices respectively. In the case of axial rod element, F_j, for instance is defined by,

$$\lfloor F_j \rfloor = \frac{L}{AE} \tag{5.7}$$

where the j denotes the jth rod flexibility matrix.

In compact matrix form, Eq. (5.6) becomes,

$$U = \tfrac{1}{2}\lfloor I \rfloor [F]\{I\}. \tag{5.6a}$$

In order to express Eq. (5.6a) in terms of the redundant forces, R_m, alone, we note from Eq. (5.5b) that,

$$\{I\} = \left[[B_0] \,\vdots\, [B_1] \right] \left\{ \begin{matrix} \{Q\} \\ \{R\} \end{matrix} \right\}, \tag{5.5b}$$

and

$$\{I\}^T = \lfloor I \rfloor = \left[\lfloor Q \rfloor \,\vdots\, \lfloor R \rfloor \right] \left[\begin{matrix} [B_0] \\ [B_I] \end{matrix} \right]. \tag{5.8}$$

Upon substituting Eqs. (5.5b) and (5.8) into Eq. (5.6a) yields the following,

$$U = \tfrac{1}{2}\left\{ \lfloor Q \rfloor \,\vdots\, \lfloor R \rfloor \right\} [D] \left\{ \begin{matrix} \{Q\} \\ \{R\} \end{matrix} \right\}, \tag{5.9}$$

where

$$[D] = \left[[B_0] \,\vdots\, [B_1] \right]^T [F] \left[[B_0] \,\vdots\, [B_1] \right] \tag{5.10}$$

Since Q and R are the applied and redundant forces, respectively, the matrix $[D]$ can be partitioned accordingly, or,

$$U = \tfrac{1}{2}\Big[\lfloor Q \rfloor \,\Big|\, \lfloor R \rfloor \Big] \left[\begin{array}{c|c} D_{QQ} & D_{QR} \\ \hline D_{RQ} & D_{RR} \end{array}\right] \left\{ \begin{array}{c} \{Q\} \\ \hline \{R\} \end{array} \right\}. \tag{5.11}$$

After expanding Eq. (5.11) becomes,

$$U = \tfrac{1}{2}\Big(\lfloor Q \rfloor [D_{QQ}]\{Q\} + \lfloor Q \rfloor [D_{QR}]\{R\} + \lfloor R \rfloor [D_{RQ}]\{Q\} + \lfloor R \rfloor [Q_{RR}]\{R\} \Big). \tag{5.12}$$

Due to the symmetry in the $[D]$ matrix, Eq. (5.12) may be written as,

$$U = \tfrac{1}{2}\Big(\lfloor Q \rfloor [D_{QQ}]\{Q\} + 2\lfloor R \rfloor [D_{RQ}]\{Q\} + \lfloor R \rfloor [D_{RR}]\{R\} \Big) \tag{5.12a}$$

Previously, it was stated that due to the compatibility conditions (continuity of displacements), the relative displacement of each member at the "cut" must vanish. Hence, utilizing Castigliano's Second Theorem and carrying out the differentiation with respect to R in Eq. (5.12a) yields,

$$[D_{RQ}]\{Q\} + [D_{RR}]\{R\} = 0. \tag{5.13}$$

Premultiplying Eq. (5.13) by the inverse of $[D_{RR}]$ and solving for $\{R\}$ the following is obtained,

$$\{R\} = -[D_{RR}]^{-1}[D_{RQ}]\{Q\} \tag{5.14}$$

Knowing $\{R\}$ from Eq. (5.14) all the internal forces can be obtained by utilizing Eq. (5.5a).

So far, the development of the theory for the force method has been limited to that of evaluating internal forces only. Since displacements, quite often are used as the criteria for structural failure, we shall extend the method in order to be able to determine such displacements. To do so, the complementary strain energy Eq. (5.12a) will be first expressed in terms of the external applied nodal forces. This by utilizing Castigliano's Second Theorem will yield the desired functional relationship between the external nodal forces and the nodal displacements.

Due to the symmetry in the $[D]$ matrix, we can easily deduce the following identity,

$$\lfloor Q \rfloor [D_{QR}]\{R\} = \lfloor R \rfloor [D_{RQ}]\{Q\}. \tag{5.15}$$

Thus, making use of Eqs. (5.14) and (5.15), Eq. (5.12a) then becomes,

$$U = \tfrac{1}{2}\Big(\lfloor Q \rfloor [D_{QQ}]\{Q\} - \lfloor Q \rfloor [D_{QR}][D_{RR}]^{-1}[D_{RQ}]\{Q\} \Big). \tag{5.16}$$

Now, applying Castigliano's Theorem yields,

$$\frac{\partial U}{\partial q} = \{q\} = [D_{QQ}]\{Q\} - [D_{QR}][D_{RR}]^{-1}[D_{RQ}]\{Q\}. \tag{5.17}$$

Factoring out $\{Q\}$ the following is obtained,

$$\{q\} = \Big[[D_{QQ}] - [D_{QR}][D_{RR}]^{-1}[D_{RQ}]\Big]\{Q\}, \tag{5.17a}$$

or

$$\{q\} = [\alpha]\{Q\}, \tag{5.17b}$$

where

$$[\alpha] = [D_{QQ}] - [D_{QR}][D_{RR}]^{-1}[D_{RQ}]. \tag{5.18}$$

The matrix, $[\alpha]$, is referred to as the influence matrix and is always symmetrical. It relates applied external nodal forces to the nodal displacements on the structure. A typical coefficient, α_{ij} for instance, is defined as the displacement at node i caused by a unit force applied at j. It is important at this point to note the difference between the flexibility matrix of the members $[F]$, and the structure influence matrix, $[\alpha]$. In the latter, nodal displacements are related to external forces, while in the former, the member displacements are related to their respective internal forces.

Summarizing, the following steps lead to the solution of structural problems by the use of the force method.

1. Find the $[B_0]$ matrix.
2. Find the $[B_1]$ matrix.
3. Find the flexibility matrix for all members.
4. Formulate the $[D]$ matrix, Eq. (5.10).
5. Find the redundant forces, Eq. (5.14).
6. Find the internal forces on each member, Eq. (5.5).
7. Formulate the $[\alpha]$ matrix, Eq. (5.18).
8. Find the nodal displacements, Eq. (5.17b).

5.3 ANALYSIS OF TRUSS-STRUCTURES BY THE FORCE MATRIX METHOD

In this section, the application of the force method to truss-type structures is presented. Simple trusses are used merely to illustrate the theory. All the solutions are carried out by hand. For larger truss-problems where the order of the matrix to be inverted exceeds 3×3, one may resort to a digital computer for a solution.

Illustrative Examples

Example 1

For purposes of illustration, let us consider the indeterminate truss-structure shown in Fig. 5.1(a). As it can be easily seen, the truss is redundant to the second degree. For the selection of the redundant members several choices exist. Here if we choose members 10 and 11 to be the redundant members then the truss can be reduced to a determinate one as shown in Fig. 5.1(b). To formulate the $[B_0]$ matrix we set unit values for Q_i $(i = 1, 2)$ one at a time with all others (Q and R) held at zero and find result internal forces on all members in each case. Therefore doing this yields,

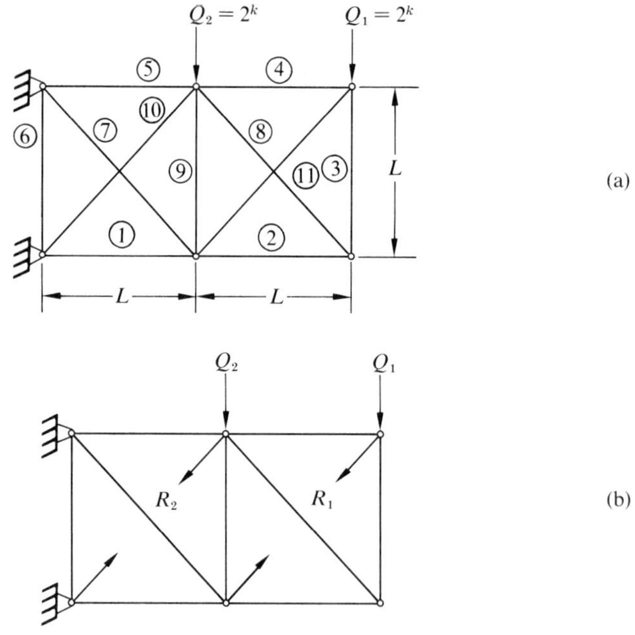

Fig. 5.1. Truss-structure.

$$[B_0] = \begin{bmatrix} -2 & -1 \\ -1 & 0 \\ -1 & 0 \\ 0 & 0 \\ 1 & 0 \\ 1.4 & 1.4 \\ 1.4 & 0 \\ -1 & -1 \\ 0 & 0 \\ 0 & 0 \end{bmatrix} \tag{a}$$

Columns: $Q_1 = 1$, $Q_2 = 1$; rows (Member): 1, 2, 3, 4, 5, 7, 8, 9, 10, 11

In the above negative signs mean members are in compression; otherwise they are in tension.

Similarly, to formulate the $[B_1]$ matrix we set unit values for R_i $(i = 1, 2)$ one at a time with all others (Q and R) held zero and calculate the result internal forces in each case. Thus,

$$
[B_1] = \begin{bmatrix} 0 & -0.72 \\ -0.71 & 0 \\ -0.71 & 0 \\ -0.71 & 0 \\ 0 & -0.71 \\ 0 & 1 \\ 1 & 0 \\ -0.71 & -0.71 \\ 0 & 1 \\ 1 & 0 \end{bmatrix}
\begin{matrix} \text{Member} \\ 1 \\ 2 \\ 3 \\ 4 \\ 5 \\ 7 \\ 8 \\ 9 \\ 10 \\ 11 \end{matrix}
\quad \text{columns: } R_1 = 1,\ R_2 = 1 \tag{b}
$$

From Eq. (5.7) the flexibility matrix for the rods is,

$$
[F] = \frac{L}{AE} \begin{bmatrix} 1 &&&&&&&&& \\ & 1 &&&&&&&& \\ && 1 &&&&&&& \\ &&& 1 &&&&&& \\ &&&& 1 &&&&& \\ &&&&& 1.4 &&&& \\ &&&&&& 1.4 &&& \\ &&&&&&& 1 && \\ &&&&&&&& 1.4 & \\ &&&&&&&&& 1.4 \end{bmatrix} \tag{c}
$$

From Eq. (5.10).

$$
[D] = \left[[B_0] \,\vdots\, [B_1] \right]^T [F] \left[[B_0] \,\vdots\, [B_1] \right]
$$

$$
= \frac{L}{AE} \left[\begin{array}{cc|cc} 13.66 & 5.8 & 4.1 & 3.4 \\ 5.8 & 4.8 & 0.71 & 3.4 \\ \hline 4.1 & 0.71 & 4.8 & 0.5 \\ 3.4 & 3.4 & 0.5 & 4.3 \end{array} \right] = \left[\begin{array}{c|c} D_{QQ} & D_{QR} \\ \hline D_{RQ} & D_{RR} \end{array} \right] \tag{d}
$$

The unknown redundants from Eq. (5.14) are.

$$
\begin{bmatrix} R_1 \\ R_2 \end{bmatrix} = - \begin{bmatrix} \dfrac{4.8L}{AE} & \dfrac{L}{2AE} \\ \dfrac{L}{2AE} & \dfrac{4.3L}{AE} \end{bmatrix}^{-1} \begin{bmatrix} \dfrac{4.1L}{AE} & \dfrac{0.71L}{AE} \\ \dfrac{3.4L}{AE} & \dfrac{3.4L}{AE} \end{bmatrix} \begin{bmatrix} Q_1 \\ Q_2 \end{bmatrix} \tag{e}
$$

For simplicity, let $Q_1 = Q_2 = Q$, solving for R_1 and R_2 yields,

$$R_1 = -1.17\,Q, \qquad R_2 = -1.44\,Q \tag{f}$$

The internal forces (Eq. 5.5) are,

$$\begin{bmatrix} I_1 \\ I_2 \\ I_3 \\ I_4 \\ I_5 \\ I_7 \\ I_8 \\ I_9 \\ I_{10} \\ I_{11} \end{bmatrix} = \begin{bmatrix} -2 & -1 \\ -1 & 0 \\ -1 & 0 \\ 0 & 0 \\ 1 & 0 \\ 1.4 & 1.4 \\ 1.4 & 0 \\ -1 & -1 \\ 0 & 0 \\ 0 & 0 \end{bmatrix} \begin{bmatrix} Q \\ Q \end{bmatrix} + \begin{bmatrix} 0 & -0.71 \\ -0.71 & 0 \\ -0.71 & 0 \\ -0.71 & 0 \\ 0 & -0.71 \\ 0 & 1 \\ 1 & 0 \\ -0.71 & -0.71 \\ 0 & 1 \\ 1 & 0 \end{bmatrix} \begin{bmatrix} -1.17\,Q \\ -1.44\,Q \end{bmatrix} \tag{g}$$

or

$$\begin{aligned} I_1 &= -1.95\,Q & I_7 &= 1.36\,Q \\ I_2 &= -0.17\,Q & I_8 &= 0.24\,Q \\ I_3 &= -0.17\,Q & I_9 &= -0.12\,Q \\ I_4 &= 0.83\,Q & I_{10} &= -1.44\,Q \\ I_5 &= 2.05\,Q & I_{11} &= -1.17\,Q \end{aligned} \tag{h}$$

In the above negative signs mean members are in compression; otherwise they are in tension.

Example 2

Figure 5.2 shows a truss-structure which has one redundancy. Assuming that AE is constant, find all member internal forces.

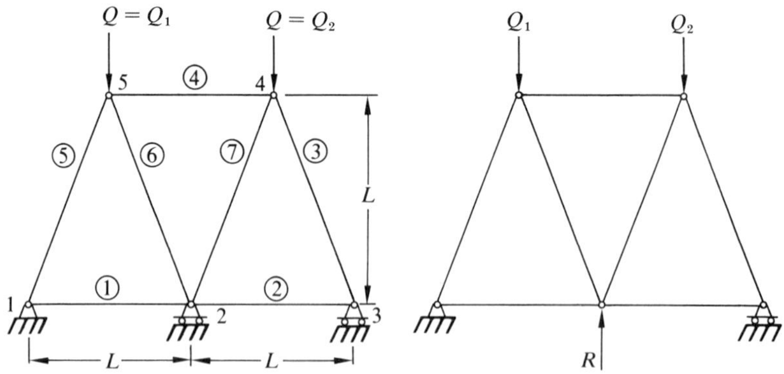

Fig. 5.2. Truss-structure.

In this example let us select the support reaction as the redundant. Other selections could have been made. Thus, removing the support at node 2 and introducing the reaction as the redundant will make the truss under consideration statically determinate. In a similar procedure as before, the $[B_0]$, $[B_1]$, and $[F]$ matrices can be obtained and are given as,

$$\begin{array}{ccccc} & Q_1=1 \quad Q_2=1 & & R=1 & \text{Member} \\ [B_0]= & \begin{bmatrix} 0.433 & 0.144 \\ 0.144 & 0.433 \\ -0.29 & -0.866 \\ -0.29 & -0.29 \\ -0.866 & -0.29 \\ -0.29 & 0.29 \\ 0.29 & -0.29 \end{bmatrix}; & [B_1]= & \begin{bmatrix} -0.29 \\ -0.29 \\ 0.58 \\ 0.58 \\ 0.58 \\ -0.58 \\ -0.58 \end{bmatrix} & \begin{matrix} 1 \\ 2 \\ 3 \\ 4 \\ 5 \\ 6 \\ 7 \end{matrix} \end{array} \qquad \text{(a)}$$

$$[F]=\frac{l}{AE}\begin{bmatrix} 1 & & & & & & \\ & 1 & & & & & \\ & & 1 & & & & \\ & & & 1 & & & \\ & & & & 1 & & \\ & & & & & 1 & \\ & & & & & & 1 \end{bmatrix} \qquad \text{(b)}$$

From Eq. (5.10),

$$[D]=\left[[B_0]\,\vdots\,[B_1]\right]^T[F]\left[[B_0]\,\vdots\,[B_1]\right],$$

$$\left[\begin{array}{cc|c} 1.21 & 0.54 & -1.52 \\ 0.54 & 1.29 & -1.02 \\ \hline -1.52 & -1.02 & 1.84 \end{array}\right]=\left[\begin{array}{c|c} D_{QQ} & D_{QR} \\ \hline D_{RQ} & D_{RR} \end{array}\right] \qquad \text{(c)}$$

From Eq. (5.14),

$$R=-[D_{RR}]^{-1}[D_{RQ}]\{Q\}.$$

Therefore,

$$R=-\frac{1}{1.84}[-1.52 \quad -1.02]\begin{bmatrix} Q \\ Q \end{bmatrix} \qquad \text{(d)}$$

$$=1.4Q$$

The internal forces, now, can be calculated from Eq. (5.5),

$$\begin{bmatrix} I_1 \\ I_2 \\ I_3 \\ I_4 \\ I_5 \\ I_6 \\ I_7 \end{bmatrix}=\begin{bmatrix} 0.433 & 0.144 \\ 0.144 & 0.433 \\ -0.29 & -0.866 \\ -0.29 & -0.29 \\ -0.866 & -0.29 \\ -0.29 & 0.29 \\ 0.29 & -0.29 \end{bmatrix}\begin{bmatrix} Q \\ Q \end{bmatrix}+\begin{bmatrix} -0.29 \\ -0.29 \\ 0.58 \\ 0.58 \\ 0.58 \\ -0.58 \\ -0.58 \end{bmatrix}[1.4Q] \qquad \text{(e)}$$

or

$$
\begin{array}{ll}
I_1 = 0.17Q & I_5 = -0.344Q \\
I_2 = 0.17Q & I_6 = -0.812Q \\
I_3 = -0.344Q & I_7 = -0.812Q \\
I_4 = 0.232Q &
\end{array}
\tag{f}
$$

Example 3

As a third example, assume that the displacements at nodes 2, 4, and 5 of the truss shown in Fig. 5.1 are required. Since node 2 is a free node (no forces exist), then, in order to find its displacement a fictitious force must be applied at that point. Assuming a fictitious force Q_3 at node 2, the $[B_0]$ matrix becomes (*see* Eq. (a) of example 1):

$$
\begin{array}{c}
\quad Q_1 = 1 \quad Q_2 = 1 \quad Q_3 = 1 \qquad \text{Member}
\end{array}
$$

$$
[B_0] = \begin{bmatrix} -2 & -1 & -1 \\ -1 & 0 & 0 \\ -1 & 0 & 0 \\ 0 & 0 & 0 \\ 1 & 0 & 0 \\ 1.4 & 1.4 & 1.4 \\ 1.4 & 0 & 0 \\ -1 & -1 & 0 \\ 0 & 0 & 0 \\ 0 & 0 & 0 \end{bmatrix} \quad \begin{array}{c} 1 \\ 2 \\ 3 \\ 4 \\ 5 \\ 7 \\ 8 \\ 9 \\ 10 \\ 11 \end{array} \tag{a}
$$

The member flexibility matrix and the $[B_1]$ stay the same and are given in Eqs. (c) and (b) in Example 2, respectively.

Forming the D matrix yields,

$$
[D] = \left[[B_0] \mid [B_1]\right]^T [F] \left[[B_0] \mid [B_1]\right],
$$

$$
\frac{L}{AE} \left[\begin{array}{ccc|cc} 13.6 & 5.8 & 4.8 & 4.13 & 3.42 \\ 5.8 & 4.8 & 3.8 & 0.71 & 3.42 \\ 4.8 & 3.8 & 3.8 & 0 & 2.71 \\ \hline 4.13 & 0.71 & 0 & 10.8 & 2 \\ 3.42 & 3.42 & 2.71 & 2 & 8.4 \end{array}\right] = \left[\begin{array}{c|c} D_{QQ} & D_{QR} \\ \hline D_{RQ} & D_{RR} \end{array}\right] \tag{b}
$$

From Eq. (5.18) the $[\alpha]$ matrix is,

$$
[\alpha] = [D_{QQ}] - [D_{QR}][D_{RR}]^{-1}[D_{RQ}],
$$

or

$$
[\alpha] = \frac{L}{AE} \begin{bmatrix} 11.2 & 4.5 & 3.9 \\ 4.5 & 3.4 & 2.7 \\ 3.9 & 2.7 & 2.9 \end{bmatrix} \tag{c}
$$

Therefore, from Eq. (5.17b) the deflections are,

$$\begin{bmatrix} q_4 \\ q_5 \\ q_2 \end{bmatrix} = \frac{L}{AE}\begin{bmatrix} 11.2 & 4.5 & 3.9 \\ 4.5 & 3.4 & 2.7 \\ 3.9 & 2.7 & 2.9 \end{bmatrix}\begin{bmatrix} Q \\ Q \\ 0 \end{bmatrix} \tag{d}$$

or

$$q_4 = \frac{15.7Q}{AE}, \qquad q_5 = \frac{7.9Q}{AE}, \qquad q_2 = \frac{6.6Q}{AE}. \tag{c}$$

Notice that the order of q must follow the order in setting up the $[B_0]$ matrix.

Example 4

Given the truss-structure shown in Fig. 5.3(a) find all the internal forces and the vertical and horizontal displacements at node 2. Assume AE to be constant for all members.

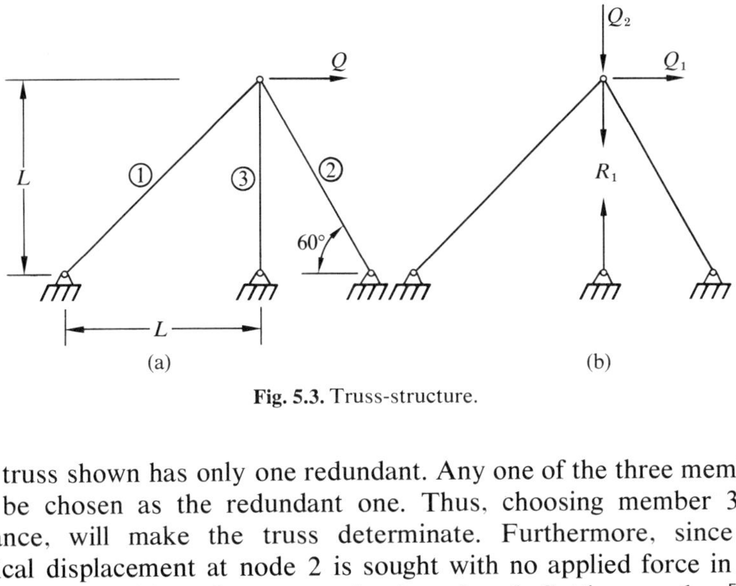

Fig. 5.3. Truss-structure.

The truss shown has only one redundant. Any one of the three members can be chosen as the redundant one. Thus, choosing member 3 for instance, will make the truss determinate. Furthermore, since the vertical displacement at node 2 is sought with no applied force in that direction, a fictitious force must be introduced. Setting up the $[B_0]$, $[B_1]$ and $[F]$ matrices yields,

$$\begin{array}{c} \\ [B_0] \end{array}\begin{array}{c} Q_1 = 1 \quad Q_2 = 1 \\ \begin{bmatrix} 0.89 & -0.52 \\ -0.73 & -0.73 \\ 0 & 0 \end{bmatrix} \end{array} \quad [B_1] = \begin{array}{c} R_1 = 1 \\ \begin{bmatrix} -0.52 \\ -0.73 \\ 1 \end{bmatrix} \end{array} \quad F = \frac{L}{AE}\begin{bmatrix} 1.4 & & \\ & 1.15 & \\ & & 1 \end{bmatrix} \quad \begin{array}{c} \text{Member} \\ 1 \\ 2 \\ 3 \end{array} \tag{a}$$

The $[D]$ matrix from Eq. (5.10),

$$[D] = \Big[[B_0] \,\vdots\, [B_1]\Big]^T [F] \Big[[B_0] \,\vdots\, [B_1]\Big]$$

$$\frac{L}{AE}\left[\begin{array}{cc|c} 1.73 & 0.03 & 0.03 \\ -0.03 & 1 & 1 \\ \hline -0.03 & 1 & 2 \end{array}\right] = \left[\begin{array}{c|c} D_{QQ} & D_{QR} \\ \hline D_{RQ} & D_{RR} \end{array}\right] \tag{b}$$

The unknown redundant from Eq. (5.14),

$$R = -[D_{RR}]^{-1}[D_{RQ}]\{Q\} = -\tfrac{1}{2}[-0.03 \quad 1]\begin{bmatrix} Q \\ 0 \end{bmatrix} = 0.015Q \tag{c}$$

Therefore, the internal loads from Eq. (5.5) are,

$$\begin{bmatrix} I_1 \\ I_2 \\ I_3 \end{bmatrix} = \begin{bmatrix} 0.89 & -0.52 \\ -0.73 & -0.73 \\ 0 & 0 \end{bmatrix}\begin{bmatrix} Q \\ 0 \end{bmatrix} + \begin{bmatrix} -0.52 \\ -0.73 \\ 1 \end{bmatrix}[0.015Q] \tag{d}$$

or

$$I_1 = 0.882Q, \qquad I_2 = -0.74Q, \qquad I_3 = 0.015Q. \tag{e}$$

To find the displacements, the influence matrix from Eq. (5.18) is,

$$\begin{aligned}[\alpha] &= [D_{QQ}] - [D_{QR}][D_{RR}]^{-1}[D_{RQ}] \\ &= \frac{L}{AE}\begin{bmatrix} 1.73 & -0.015 \\ -0.015 & 0.5 \end{bmatrix}\end{aligned} \tag{f}$$

Therefore, the displacements from Eq. (5.17b) are,

$$\begin{bmatrix} q_1 = \delta_2^{\,x} \\ q_2 = \delta_2^{\,y} \end{bmatrix} \frac{L}{AE}\begin{bmatrix} 1.73 & 0.015 \\ -0.015 & 0.5 \end{bmatrix}\begin{bmatrix} Q \\ 0 \end{bmatrix} \tag{g}$$

or

$$\delta_2^{\,x} = \frac{1.73\,QL}{AE}, \qquad \delta_2^{\,y} = -\frac{0.015\,QL}{AE}. \tag{h}$$

5.4 ANALYSIS OF RIGID FRAME AND BEAM STRUCTURES BY THE FORCE MATRIX METHOD

The application of the force method in the solution of rigid frame and beam type problems is presented in this section. The frames of reference considered in this text for the internal forces are shown in Fig. 5.4. For the system of forces as in Fig. 5.4(a), the flexibility matrix will relate the displacements, $\delta_i^{\,y}$ and θ_i to the force system V_i and M_i. On the other hand, the flexibility matrix in reference to Fig. 5.4(b), relates the displacements θ_i and θ_j to the force system M_i and M_j.

Fig. 5.4. Force method notation for element forces on beam elements.

To develop the first type, let us express the bending moment and shear at any section x distance from node i in terms of M_i and V_i,

$$V(x) = V_i, \tag{5.19a}$$

$$M(x) = V_i x + M_i. \tag{5.19b}$$

The strain energy, considering bending only, is,

$$U = \tfrac{1}{2}\int_0^L \frac{M^2}{EI}dx \tag{5.20}$$

Substituting Eq. (5.19b) into Eq. (5.20) and integrating yield,

$$U = \tfrac{1}{2}EI\left(\frac{V_i^2L^3}{3} + M_iV_iL^2 + M_i^2L\right). \tag{5.21}$$

Utilizing Castigliano's Second Theorem, the displacements at node i are obtained,

$$\frac{\partial U}{\partial V_i} = \delta_i^y = \frac{L^3}{3EI}V_i + \frac{L^2}{2EI}M_i \tag{5.22a}$$

$$\frac{\partial U}{\partial M_i} = \theta_i = \frac{L^2}{2EI}V_i + \frac{L}{EI}M_i \tag{5.22b}$$

In matrix notation Eqs. (5.22a) and (5.22b) can be expressed as,

$$\begin{bmatrix} \delta_i^y \\ \theta_i \end{bmatrix} = \frac{1}{EI}\begin{bmatrix} L^3/3 & L^2/2 \\ L^2/2 & L \end{bmatrix}\begin{bmatrix} V_i \\ M_i \end{bmatrix} \tag{5.22c}$$

Where type (a) beam flexibility matrix now is,

$$[F_i]^{(a)} = \begin{bmatrix} L^3/3EI & L^2/2EI \\ L^2/2EI & L/EI \end{bmatrix}_i \tag{5.23}$$

To find type (b) beam flexibility matrix, from equilibrium (*see* Fig. 5.4) the following is obtained,

$$M_j = M_i + V_iL, \tag{5.24}$$

or solving for V_i yields.

$$V_i = \frac{M_j - M_i}{L}. \tag{5.24a}$$

Upon substituting Eq. (5.24a) into Eq. (5.19b), the result into Eq. (5.20), integrating, and then using Castigliano's Second Theorem, results in,

$$\begin{bmatrix} \theta_i \\ \theta_j \end{bmatrix} = \frac{L}{6EI}\begin{bmatrix} 2 & 1 \\ 1 & 2 \end{bmatrix}\begin{bmatrix} M_i \\ M_j \end{bmatrix} \tag{5.25}$$

Where type (b) beam flexibility matrix is,

$$[F_i]^{(b)} = \frac{L}{6EI}\begin{bmatrix} 2 & 1 \\ 1 & 2 \end{bmatrix}_i \tag{5.26}$$

The application of the force method, and the use of Eqs. (5.23) and (5.26) in the solution of beam and frame structures can be best illustrated by considering the following examples.

Example 5 (Type A Solution)

Figure 5.5 shows a beam structure which is redundant to the second degree. The vertical displacements at node 2 and all the internal forces are required. Assume that the bending rigidity, EI, is constant.

For the selection of the redundant forces, several choices exist. Here, let us choose the reactions at node 3 to be the redundants. Other choices could have been considered.

To find the $[B_0]$ matrix, unit force $Q_1 = 1$ corresponding to P is applied, and the shears and moments at the left end of each segment are cal-

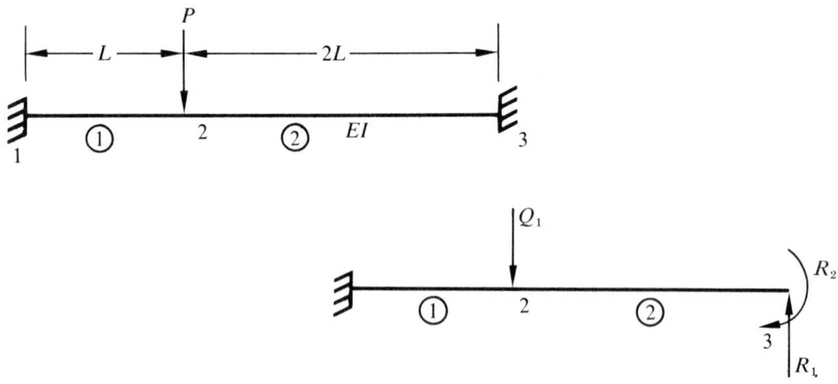

Fig. 5.5. Beam structure.

culated. Thus performing the calculation yields,

$$Q_1 = 1$$

$$[B_0] = \begin{bmatrix} 1 \\ -L \\ 0 \\ 0 \end{bmatrix} \begin{matrix} V_1 \\ M_1 \\ V_2 \\ M_2 \end{matrix} \tag{a}$$

where V_i and M_i $(i = 1, 2)$ are the shear and moment respectively, at the left section of beam i. Likewise, applying unit forces R_i $(i = 1, 2)$ one at a time and finding internal forces in each case yield,

$$R_1 = 1 \quad R_2 = 1$$

$$[B_1] \begin{bmatrix} -1 & 0 \\ 3L & -1 \\ -1 & 0 \\ 2L & -1 \end{bmatrix} \begin{matrix} V_1 \\ M_1 \\ V_2 \\ M_2 \end{matrix} \tag{b}$$

Note that the sign convention for the forces at the left section of each beam follows the sign convention adopted in Fig. 5.4(a).

From Eq. (5.23) the member flexibility matrix is,

$$[F]^{(a)} = \frac{L}{6EI} \begin{bmatrix} \begin{bmatrix} 2L^2 & 2L \\ 3L & 6 \end{bmatrix}^{(a)} & \\ & \begin{bmatrix} 16L^2 & 12L \\ 12L & 12 \end{bmatrix}^{(a)} \end{bmatrix} \tag{c}$$

The $[D]$ matrix from Eq. (5.10) is,

$$[D] = \frac{L}{6EI} \left[\begin{array}{c|cc} 2L^2 & -8L^2 & 3L \\ \hline -8L^2 & 54L^2 & -27L \\ 3L & -27L & 18 \end{array}\right] = \left[\begin{array}{c|c} D_{QQ} & D_{QR} \\ \hline D_{RQ} & D_{RR} \end{array}\right] \tag{d}$$

Therefore the redundant forces from Eq. (5.14) are,

$$R = -[D_{RR}]^{-1}[D_{RQ}]\{Q\}$$

$$\begin{bmatrix} R_1 = F_3{}^y \\ R_2 = M_3 \end{bmatrix} = \begin{bmatrix} -2/27L^2 & -1/9L \\ -1/9L & -2/9 \end{bmatrix} \begin{bmatrix} -8L^2 \\ 3L \end{bmatrix} \begin{bmatrix} P \\ 0 \end{bmatrix} \tag{e}$$

or

$$F_3{}^y = \frac{7P}{27}, \qquad M_3 = \frac{2PL}{9}. \tag{f}$$

To find the displacement, the influence matrix from Eq. (5.18) is,

$$[\alpha] = \left[\frac{8L^3}{81EI}\right]. \tag{g}$$

Hence, the displacement from Eq. (5.17b) is,

$$q_1 = \delta_2{}^y = \frac{8L^3P}{81EI}. \tag{h}$$

Example 6 (Type B Solution)

Consider the beam structure given in Example 5. In this case, only the bending moments at each end of each beam element are calculated. The procedure of finding the $[B_0]$ and $[B_1]$ matrices is exactly the same as in the previous example. Thus, applying a unit external force and calculating internal bending moments yields,

$$\begin{array}{c} Q_1 = 1 \end{array}$$

$$[B_0] = \begin{bmatrix} -L \\ 0 \\ 0 \\ 0 \end{bmatrix} \begin{array}{l} M_1 \\ M_{2L} \\ M_{2R} \\ M_3 \end{array} \tag{a}$$

In the above M_{2L} and M_{2R} are the bending moments to the left and right of node 2 and correspond to beam elements 1 and 2, respectively.

Similarly the $[B_1]$ matrix is,

$$\begin{array}{cc} R_1 = 1 & R_2 = 1 \end{array}$$

$$[B_1] = \begin{bmatrix} 3L & -1 \\ 2L & -1 \\ 2L & -1 \\ 0 & -1 \end{bmatrix} \begin{array}{l} M_1 \\ M_{2L} \\ M_{2R} \\ M_3 \end{array} \tag{b}$$

From Eq. (5.26) the member flexibility matrix is,

$$[F]^{(b)} = \frac{L}{6EI} \begin{bmatrix} \begin{bmatrix} 2 & 1 \\ 1 & 2 \end{bmatrix}^{(b)} & \\ & \begin{bmatrix} 4 & 2 \\ 2 & 4 \end{bmatrix}^{(b)} \end{bmatrix} \tag{c}$$

The $[D]$ matrix from Eq. (5.10) is,

$$[D] = \frac{L}{6EI} \left[\begin{array}{c|cc} 2L^2 & -8L^2 & 3L \\ \hline -8L^2 & 54L^2 & -27L \\ 3L & -27L & 18 \end{array} \right] = \left[\begin{array}{c|c} D_{QQ} & D_{QR} \\ \hline D_{RQ} & D_{RR} \end{array} \right] \tag{d}$$

Solving for the redundant forces from Eq. (5.14) will yield exactly those obtained in the previous example.

Example 7

Shown in Fig. 5.6 is a beam structure on three supports. Assuming EI to be constant, find all the internal forces.

It can be easily seen that if any one of the three supports is removed the structure becomes a statically determinate one. Thus, removing for instance the support at node 5 and introducing the reaction as the redundant, the following is obtained:

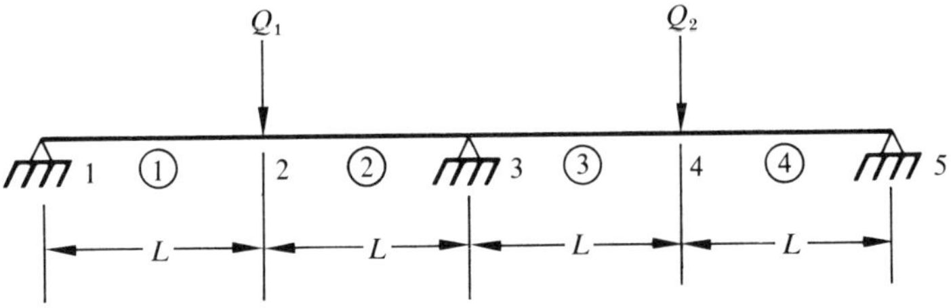

Fig. 5.6. Continuous beam structure.

Solution by type A

$$[B_0] = \begin{bmatrix} 0.5 & -0.5 \\ 0 & 0 \\ -0.5 & -0.5 \\ 0.5L & -0.5L \\ 0 & 1 \\ 0 & -L \\ 0 & 0 \\ 0 & 0 \end{bmatrix}_{Q_1=1,\ Q_2=1}, \qquad [B_1] = \begin{bmatrix} 1 \\ 0 \\ 1 \\ L \\ -1 \\ 2L \\ -1 \\ L \end{bmatrix}_{R_1=1} \begin{matrix} V_1 \\ M_1 \\ V_2 \\ M_2 \\ V_3 \\ M_3 \\ V_4 \\ M_4 \end{matrix} \tag{a}$$

The member flexibility matrix from Eq. (5.23) is.

$$[F]^{(a)} \frac{L}{6EI} \begin{bmatrix} \begin{bmatrix} 2L^2 & 3L \\ 3L & 6 \end{bmatrix}^{(a)} & & & \\ & \begin{bmatrix} 2L^2 & 3L \\ 3L & 6 \end{bmatrix}^{(a)} & & \\ & & \begin{bmatrix} 2L^2 & 3L \\ 3L & 6 \end{bmatrix}^{(a)} & \\ & & & \begin{bmatrix} 2L^2 & 3L \\ 3L & 6 \end{bmatrix}^{(a)} \end{bmatrix} \tag{b}$$

Solving for the $[D]$ matrix yield,

$$[D]=\frac{L}{6EI}\left[\begin{array}{cc|c} L^2 & -3L^2/2 & 3L^2 \\ -3L^2/2 & 6L^2 & -13L^2 \\ \hline 3L^2 & -13L^2 & 32L^2 \end{array}\right]=\left[\begin{array}{c|c} D_{QQ} & D_{QR} \\ \hline D_{RQ} & D_{RR} \end{array}\right] \tag{c}$$

Hence, the redundant from Eq. (5.14), assuming for simplicity that $Q_1=P$ and $Q_2=0$, is,

$$R_1=F_5^{\,y}=-3P/32 \tag{d}$$

The internal forces from Eq. (5.5) are,

$$\begin{bmatrix} V_1 \\ M_1 \\ V_2 \\ M_2 \\ V_3 \\ M_3 \\ V_4 \\ M_4 \end{bmatrix}=\begin{bmatrix} 0.5 & -0.5 \\ 0 & 0 \\ -0.5 & -0.5 \\ 0.5L & -0.5L \\ 0 & 1 \\ 0 & -L \\ 0 & 0 \\ 0 & 0 \end{bmatrix}\begin{bmatrix} P \\ 0 \end{bmatrix}+\begin{bmatrix} 1 \\ 0 \\ 1 \\ L \\ -1 \\ 2L \\ -1 \\ L \end{bmatrix}[-3P/32] \tag{e}$$

Solution by type B

Selecting the same redundant, and finding the $[B_0]$ and $[B_1]$ matrices yield,

$$\begin{array}{ccc} & Q_1=1 \quad Q_2=1 & & & R_1=1 \\ [B_0]= & \begin{bmatrix} 0 & 0 \\ L/2 & -L/2 \\ L/2 & -L/2 \\ 0 & -L \\ 0 & -L \\ 0 & 0 \\ 0 & 0 \\ 0 & 0 \end{bmatrix} & \begin{matrix} M_1 \\ M_{2L} \\ M_{2R} \\ M_{3L}; \\ M_{3R} \\ M_{4L} \\ M_{4R} \\ M_5 \end{matrix} & [B_1] & \begin{bmatrix} 0 \\ L \\ L \\ 2L \\ 2L \\ L \\ L \\ 0 \end{bmatrix} \end{array} \tag{a}$$

From Eq. (5.26) the member flexibility matrix is given by,

$$[F]^{(b)}=\frac{L}{6EI}\begin{bmatrix} \begin{bmatrix} 2 & 1 \\ 1 & 2 \end{bmatrix}^{(b)} & & & \\ & \begin{bmatrix} 2 & 1 \\ 1 & 2 \end{bmatrix}^{(b)} & & \\ & & \begin{bmatrix} 2 & 1 \\ 1 & 2 \end{bmatrix}^{(b)} & \\ & & & \begin{bmatrix} 2 & 1 \\ 1 & 2 \end{bmatrix}^{(b)} \end{bmatrix} \tag{b}$$

Solving Eq. (5.10) for the $[D]$ matrix yields,

$$[D] = \frac{L}{6EI}\left[\begin{array}{cc|c} L^2 & 3L^2/2 & 3L^2 \\ -3L^2/2 & 6L^2 & -13L^2 \\ \hline 3L^2 & -13L^2 & 32L^2 \end{array}\right] = \left[\begin{array}{c|c} D_{QQ} & D_{QR} \\ \hline D_{RQ} & D_{RR} \end{array}\right]$$

Hence, the redundant force from Eq. (5.14) is,

$$R_1 = F_5^y = -3P/32, \tag{d}$$

which is the same as in solution type A.

The internal forces from Eq. (5.5) are given by,

$$\begin{bmatrix} M_1 \\ M_{2L} \\ M_{2R} \\ M_{3L} \\ M_{3R} \\ M_{4L} \\ M_{4R} \\ M_5 \end{bmatrix} = \begin{bmatrix} 0 & 0 \\ L/2 & -L/2 \\ L/2 & -L/2 \\ 0 & -L \\ 0 & -L \\ 0 & 0 \\ 0 & 0 \\ 0 & 0 \end{bmatrix} \begin{bmatrix} P_1 \\ 0_2 \end{bmatrix} + \begin{bmatrix} 0 \\ L \\ L \\ 2L \\ 2L \\ L \\ L \\ 0 \end{bmatrix} [-3P/32] \tag{e}$$

Example 8

Figure 5.7 shows a rigid frame structure which is statically indeterminate to the third degree. The bending rigidity (EI) is assumed to be constant for all beam elements. Also, the axial and transverse shear deformations are assumed to be negligible. Using type B solution, find all the internal forces.

Selecting the reactions at node 5 as the redundants and using the same

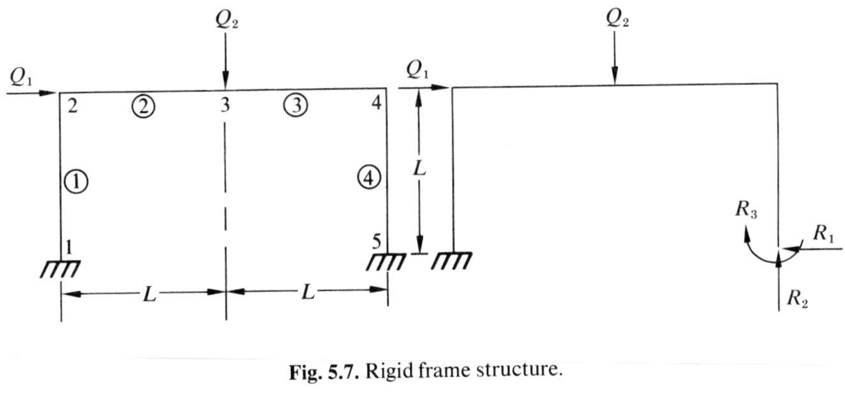

Fig. 5.7. Rigid frame structure.

procedure as before, the following can be obtained,

$$[B_0]=\begin{array}{c}\begin{array}{cc} Q_1=1 & \quad Q_2=1\end{array}\\ \begin{bmatrix} L & L\\ 0 & L\\ 0 & L\\ 0 & 0\\ 0 & 0\\ 0 & 0\\ 0 & 0\\ 0 & 0\end{bmatrix}\end{array}\begin{array}{l} \\ M_1\\ M_{2L}\\ M_{2R}\\ M3_L;\\ M_{3R}\\ M_{4L}\\ M_{4R}\\ M_5\end{array}[B_1]=\begin{array}{c}\begin{array}{ccc} R_1=1 & R_2=1 & R_3=1\end{array}\\ \begin{bmatrix} 0 & -2L & 1\\ L & -2L & 1\\ L & -2L & 1\\ L & -L & 1\\ L & -L & 1\\ L & 0 & 1\\ L & 0 & 1\\ 0 & 0 & 1\end{bmatrix}\end{array}\tag{a}$$

Notice that Eq. (a) counterclockwise moments are considered to be positive. This is arbitrary as long as the sign convention is kept consistent.

The member flexibility matrix from Eq. (5.26) is,

$$[F]^{(b)}=\frac{L}{6EI}\begin{bmatrix} \begin{bmatrix}2 & 1\\ 1 & 2\end{bmatrix}_1^{(b)} & & & \\ & \begin{bmatrix}2 & 1\\ 1 & 2\end{bmatrix}_2^{(b)} & & \\ & & \begin{bmatrix}2 & 1\\ 1 & 2\end{bmatrix}_3^{(b)} & \\ & & & \begin{bmatrix}2 & 1\\ 1 & 2\end{bmatrix}_4^{(b)}\end{bmatrix}\tag{b}$$

From Eq. (5.10) the $[D]$ is,

$$[D]=\frac{L}{6EI}\left[\begin{array}{cc|ccc} 2L^2 & 3L^2 & L^2 & -6L^2 & 3L\\ 3L^2 & 8L^2 & 6L^2 & -17L^2 & 9L\\ \hline L^2 & 6L^2 & 16L^2 & -18L^2 & 18L\\ -6L^2 & -17L^2 & -18L^2 & 40L^2 & -24L^2\\ 3L & 9L & 18L & -24L & 24\end{array}\right]=\left[\begin{array}{c|c} D_{QQ} & D_{QR}\\ \hline D_{RQ} & D_{RR}\end{array}\right]\tag{c}$$

Hence, the redundant forces from Eq. (5.14) are,

$$\begin{bmatrix} R_1=F_5{}^x\\ R_2=F_5{}^y\\ R_3=M_5\end{bmatrix}=-\frac{1}{960L^4}\begin{bmatrix} 384L^2 & 0 & -288L^3\\ 0 & 60L^2 & 60L^3\\ -288L^3 & 60L^3 & 316L^4\end{bmatrix}\begin{bmatrix} L^2 & 6L^2\\ -6L^2 & -17L^2\\ 3L & 9L\end{bmatrix}\begin{bmatrix} Q_1\\ Q_2\end{bmatrix}$$

or

$$F_5{}^x=Q_1/2+3Q_2/10$$

$$F_5{}^y=3Q_1/16+Q_2/2$$

$$M_5=-5Q_1L/16-Q_2L/10$$

For simplicity, let us assume that $Q_1 = 0$, thus the internal forces from Eq. (5.5) are,

$$\begin{bmatrix} M_1 \\ M_{2L} \\ M_{2R} \\ M_{3L} \\ M_{3R} \\ M_{4L} \\ M_{4R} \\ M_5 \end{bmatrix} = \begin{bmatrix} L \\ L \\ L \\ 0 \\ 0 \\ 0 \\ 0 \\ 0 \end{bmatrix} [Q_2] + \begin{bmatrix} 0 & -2L & 1 \\ L & -2L & 1 \\ L & -2L & 1 \\ L & -L & 1 \\ L & -L & 1 \\ L & 0 & 1 \\ L & 0 & 1 \\ 0 & 0 & 1 \end{bmatrix} \begin{bmatrix} 3Q_2/10 \\ Q_2/2 \\ -Q_2L/10 \end{bmatrix} \tag{d}$$

5.5 SUPPLEMENTARY PROBLEMS

Problem 5.1. Using the force method, find internal forces on all members for the truss-structure shown. Assume AE to be constant.

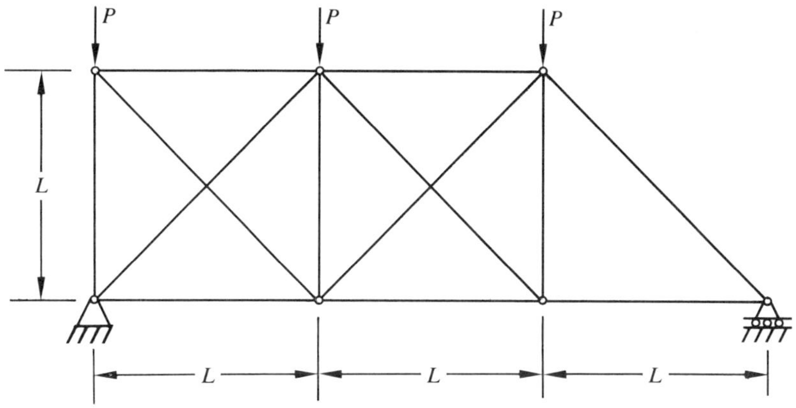

Problem 5.2. Determine the tip displacements on the indeterminate truss-structure shown. Use the force method and assume AE to be constant.

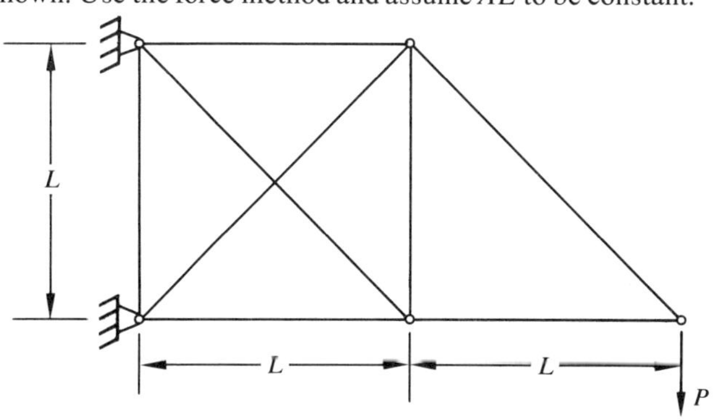

Problem 5.3. Determine all internal forces by the force method. Consider AE to be the same for all truss-members.

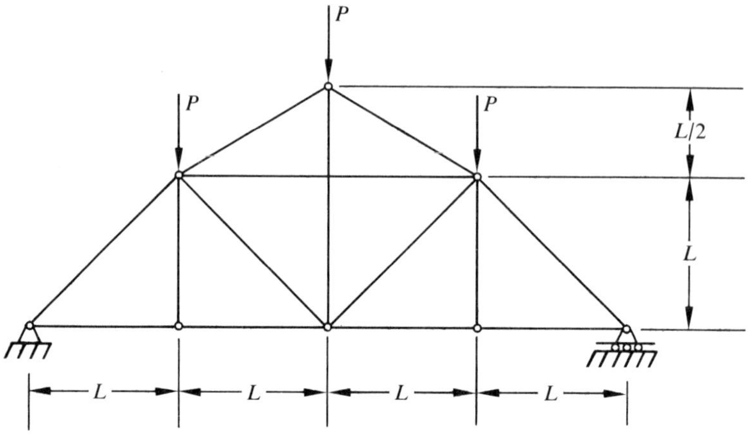

Problem 5.4. Find the rotations on all joints of the beam structures shown. Use the force method. Assume EI to be constant.

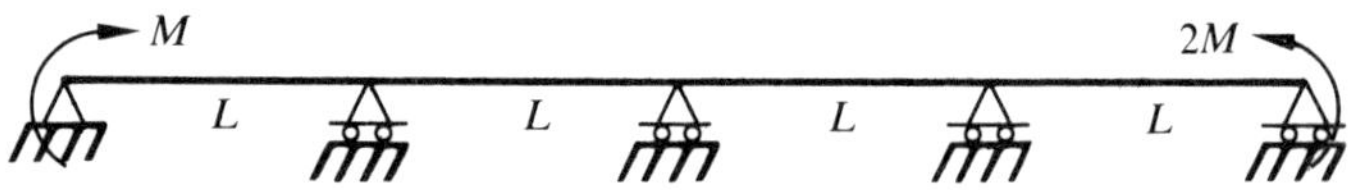

(a)

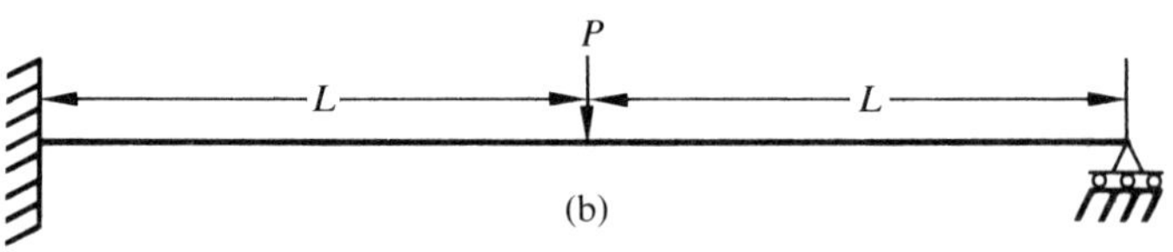

(b)

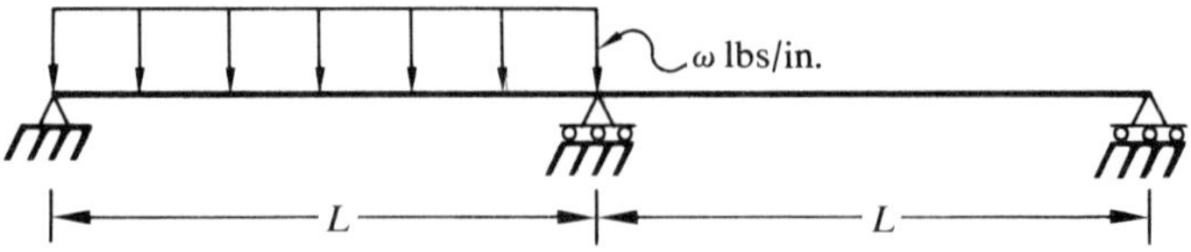

(c)

Problem 5.5. Find the influence matrix of the frame structure shown. Assume EI to be constant.

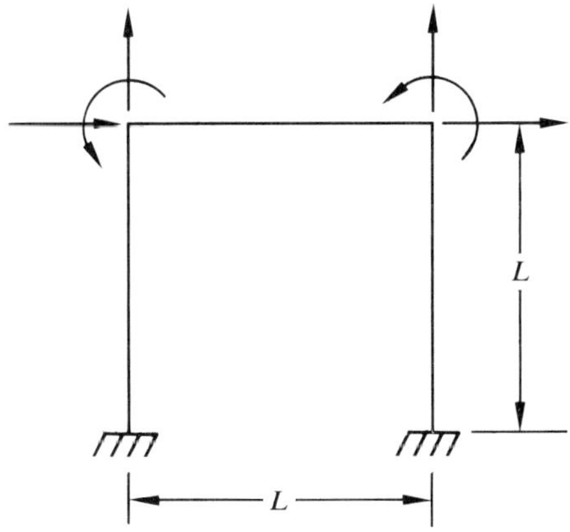

Problem 5.6. Determine the internal forces and the vertical displacement at O. Use the force method and consider EI to be constant.

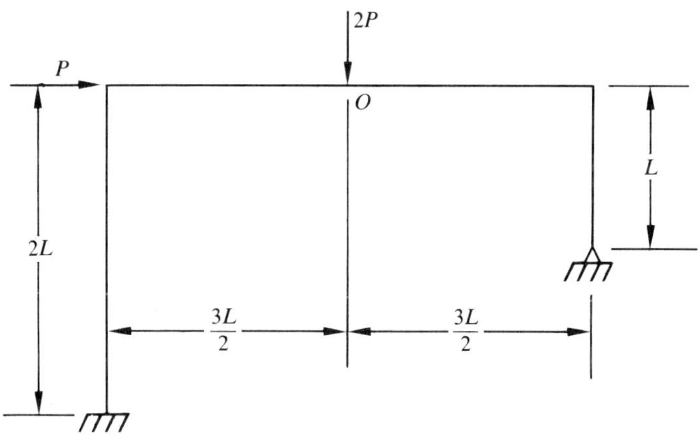

Problem 5.7. Determine the displacements (rotation and translation) of point (a) on the frame structure given below. Assume EI to be constant for both beams.

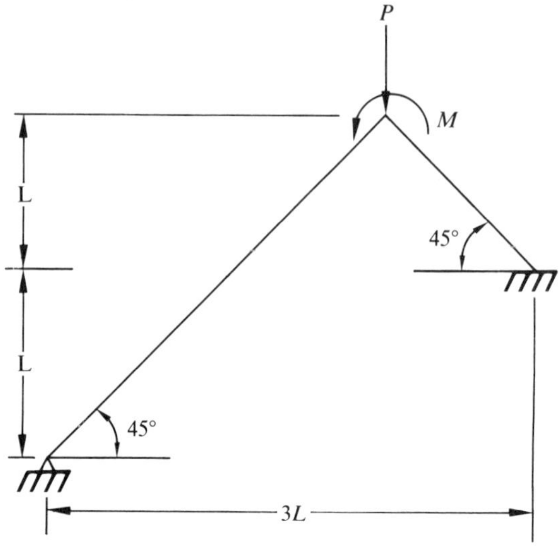

Problem 5.8. Derive the flexibility matrix for a beam element including shear deformations in addition to bending deformation. Consider two-dimensional beam element only.

Problem 5.9. Derive the flexibility matrix for a three-dimensional beam element.

Problem 5.10. Find the displacements and the internal forces on the space truss-structure shown. Use both the Matrix Force method and the Stiffness Matrix method. Compare results. Assume AE to be constant for all members.

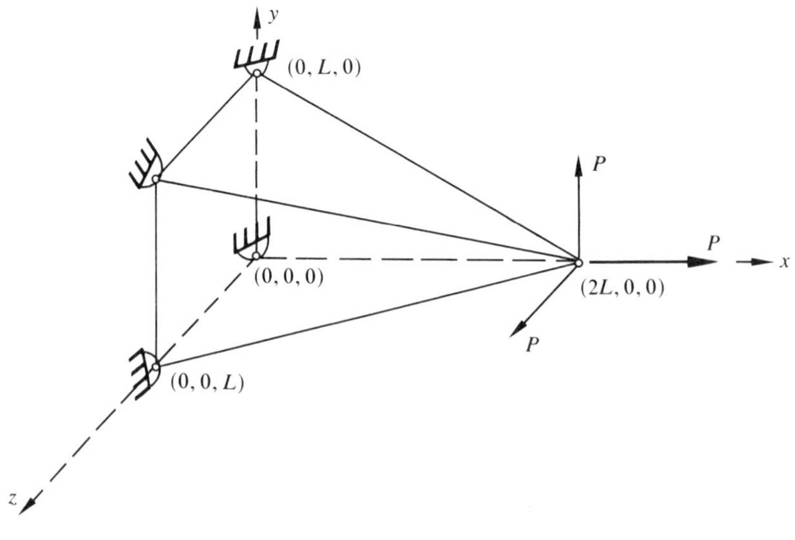

CHAPTER SIX

Analysis of Nonlinear Truss-Structures by the Stiffness Method*

6.1 INTRODUCTION

The stiffness matrix method, as presented in Chapter 3, is based upon a linear relationship between stress and strain (force and displacement) and, therefore, is not directly applicable to the analysis of structures in which the stress in any member is above the proportional limit. Since displacements are sometimes used as the criteria for failure instead of stresses, it is desirable to use a more generalized concept of the stiffness matrix method, in which the restriction on nonlinearities can be eliminated.

The matrix analysis of redundant structures with members stressed above the proportional limit has been treated in various manners. However, these methods often employ a stress-strain curve idealization which, in some cases, does not adequately represent the behavior of the material. In References [4], [9], and [28], the methods depend upon expressing the stress-strain curve by an equation of the type suggested by Ramberg and Osgood[21]. Poppleton's analysis[17] requires the idealization of the stress-strain curve into two linear segments. As shown in Figs. 6.1 and 6.2, the idealization is acceptable for aluminum and some steels, but are inadequate for mild steels used for the construction of many truss-structures. The "constant strain" method of analysis presented by Jensen *et al.*[11] is probably the most acceptable of the existing methods. The solution of the stresses and strains is slightly similar to the method presented in this chapter. However, for truss-structures, the present

*Portion of this chapter reprinted by permission of the American Society of Civil Engineers, from R. C. Hensley and J. J. Azar's Computer analysis of non-linear truss-structures in the *Journal of Structural Eng. Div.*, ASCE **94**(ST6), 1968.

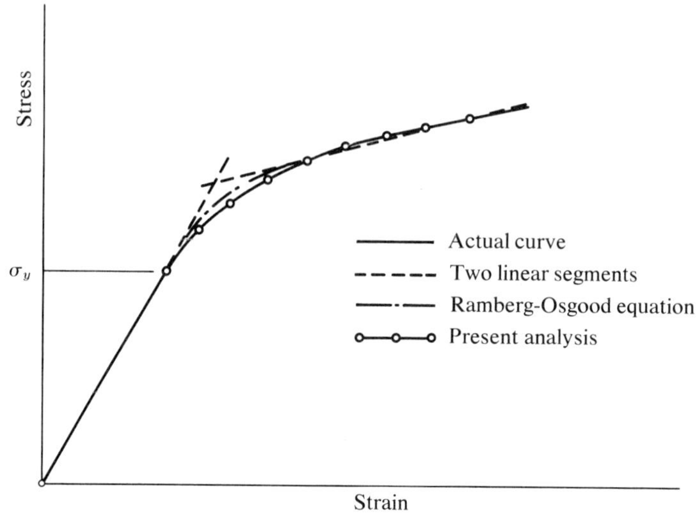

Fig. 6.1. Stress-strain curve for aluminum and some steels.

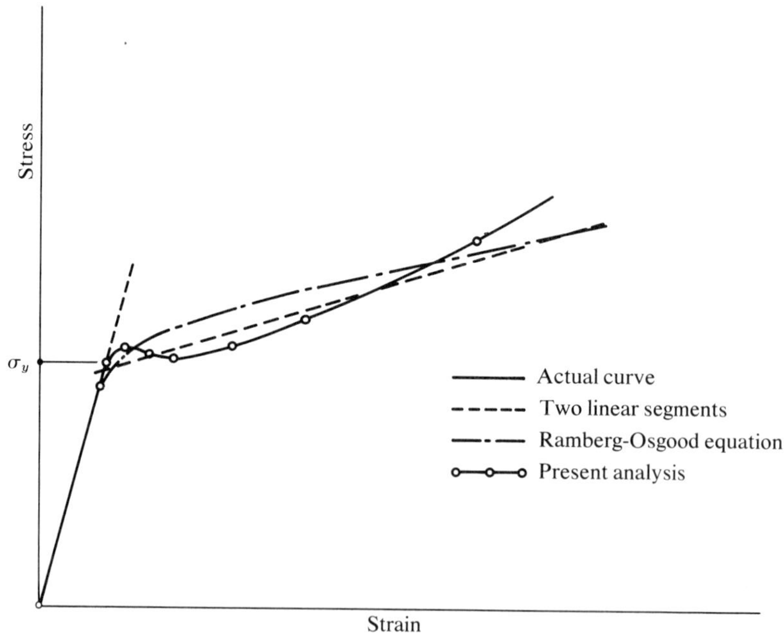

Fig. 6.2. Stress-strain curve for mild steel.

method, which uses the fictitious force technique in conjunction with the stiffness matrix method, is more concise and the data input for digital computations is greatly simplified.

6.2 MODIFIED STIFFNESS MATRIX METHOD FOR THE NONLINEAR ANALYSIS OF TRUSS-TYPE STRUCTURES

From Chapter 3, the relationship between a nodal force vector $\{Q\}$ and a nodal displacement vector $\{q\}$ for a structural element is,

$$\{Q\} = [S]\{q\}, \tag{6.1}$$

where $[S]$ is the element stiffness matrix.

For an axial rod element (Fig. 6.3) Eq. (6.1) in an expanded form becomes,

$$\begin{bmatrix} F_i^x \\ F_i^y \\ F_i^z \\ F_j^x \\ F_j^y \\ F_j^z \end{bmatrix} = \frac{AE}{L} \begin{bmatrix} R_{11}^2 & & & & \text{Sym.} & \\ R_{11}R_{12} & R_{12}^2 & & & & \\ R_{11}R_{13} & R_{12}R_{13} & R_{13}^2 & & & \\ -R_{11}^2 & -R_{11}R_{12} & -R_{11}R_{13} & R_{11}^2 & & \\ -R_{11}R_{12} & -R_{12}^2 & -R_{12}R_{13} & R_{11}R_{12} & R_{12}^2 & \\ -R_{11}R_{13} & -R_{12}R_{13} & -R_{13}^2 & R_{11}R_{13} & R_{12}R_{13} & R_{13}^2 \end{bmatrix} \begin{bmatrix} \delta_i^x \\ \delta_i^y \\ \delta_i^z \\ \delta_j^x \\ \delta_j^y \\ \delta_j^z \end{bmatrix} \tag{6.2}$$

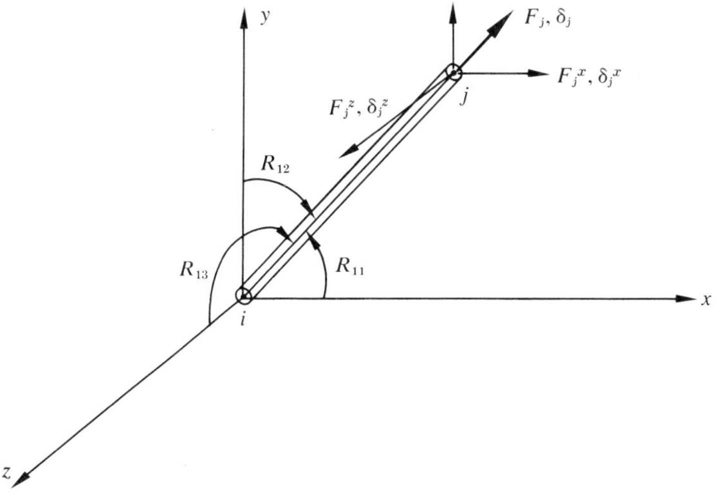

Fig. 6.3. Axial rod element.

where

A = rod cross-sectional area,
L = rod length,
E = modulus of elasticity,
R_{ij} = direction cosines,
F_i^x, F_i^y, F_j^x, etc. = internal nodal forces acting at nodes i and j of the rod.

The direction cosines may be expressed in terms of the nodal coordinates as follows,

$$R_{11} = x_{ji}/L, \qquad R_{12} = y_{ji}/L, \qquad R_{13} = z_{ji}/L, \tag{6.3}$$

where

$$L = (x_{ji}^2 + y_{ji}^2 + z_{ji}^2)^{1/2} \qquad \text{and} \qquad x_{ji} = x_j - x_i, \text{ etc.}$$

The stiffness matrix for an entire structure (referred to previously as the aggregate stiffness matrix) is formed by superposing at each node the stiffness contributions for elements connected at that node. Detailed information on this is presented in Chapter 3. Thus, the functional relationship between the externally applied nodal forces $\{F\}$, and the nodal displacements, $\{\Delta\}$ for the entire structure may be written as,

$$\{F\} = [K]\{\Delta\} \qquad \text{or} \qquad \{\Delta\} = [K]^{-1}\{F\}. \tag{6.4}$$

in which $[K]$ now is the reduced aggregate stiffness matrix and $[K]^{-1}$ is its inverse.

For the purpose of nonlinear analysis, it is desirable to formulate the internal stresses for each rod element in the structure in terms of the externally applied nodal forces. This may be easily accomplished once the nodal displacements are computed. The internal stress, σ_r, for a particular rod (r) can be expressed in the following manner,

$$\sigma_r = E_r \epsilon_r, \tag{6.5}$$

where ϵ_r is the total extensional strain in rod (r) and can be defined in terms of the rod nodal displacements as,

$$\epsilon_r = \frac{\text{change in length}}{\text{original length}} = \frac{\Delta L_r}{L_r} = \frac{(\delta_j - \delta_i)_r}{L_r}. \tag{6.6}$$

Upon substituting Eq. (6.6) into Eq. (6.5), the following is obtained,

$$\sigma_r = \frac{E_r}{L_r}(\delta_j - \delta_i)_r = \frac{E_r}{L_r}[-1 \quad 1]\begin{bmatrix} \delta_i \\ \delta_j \end{bmatrix}. \tag{6.7}$$

The displacements (δ_j and δ_i) can be expressed in terms of the reference coordinate displacements (δ_j^x, δ_j^y, δ_j^z and δ_i^x, δ_i^y, δ_i^z), respectively, as follows,

$$\begin{bmatrix} \delta_i \\ \delta_j \end{bmatrix} = \begin{bmatrix} R_{11} & R_{12} & R_{13} & 0 & 0 & 0 \\ 0 & 0 & 0 & R_{11} & R_{12} & R_{13} \end{bmatrix} \begin{bmatrix} \delta_i^x \\ \delta_i^y \\ \delta_i^z \\ \delta_j^x \\ \delta_j^y \\ \delta_j^z \end{bmatrix} \tag{6.8}$$

Substituting Eq. (6.8) into Eq. (6.7) yields,

$$\sigma_r = [-h_1 \quad -h_2 \quad -h_3 \quad h_1 \quad h_2 \quad h_3] \begin{bmatrix} \delta_i^x \\ \delta_i^y \\ \delta_i^z \\ \delta_j^x \\ \delta_j^y \\ \delta_j^z \end{bmatrix} \tag{6.9}$$

in which h_i $(i = 1, 2, 3)$ is the stress in the rod due to a unit coordinate displacement, and is,

$$h_1 = R_{11}\frac{E}{L_r}, \qquad h_2 = R_{12}\frac{E}{L_r}, \qquad h_3 = R_{13}\frac{E}{L_r}.$$

For an entire structure which is made of an assemblage of rods, the aggregate stress matrix equation can be written as,

$$\{\Sigma\} = [H]\{\Delta\}, \tag{6.10}$$

where $[H]$ is defined as the aggregate "stress-displacement" matrix.

Upon substituting Eq. (6.4) into Eq. (6.10), the following is obtained,

$$\{\Sigma\} = [H][K]^{-1}\{F\} = [\Pi]\{F\} \tag{6.11}$$

The matrix $[\Pi]$ is designated as the aggregate "stress force" matrix of which an element Π_{mn} gives the stress in the mth member due to a unit external applied force at the nth node coordinate. In the case of coplanar trusses (two-dimensional), Fig. 6.4, the element stiffness matrix Eq. (6.2) becomes,

$$\begin{bmatrix} F_i^x \\ F_i^y \\ F_j^x \\ F_j^y \end{bmatrix} = \frac{AE}{L} \begin{bmatrix} R_{11}^2 & & & \\ R_{11}R_{12} & R_{12}^2 & & \\ -R_{11}^2 & -R_{11}R_{12} & R_{11}^2 & \\ -R_{11}R_{12} & -R_{12}^2 & R_{11}R_{12} & R_{12}^2 \end{bmatrix} \begin{bmatrix} \delta_i^x \\ \delta_i^y \\ \delta_j^x \\ \delta_j^y \end{bmatrix} \tag{6.12}$$

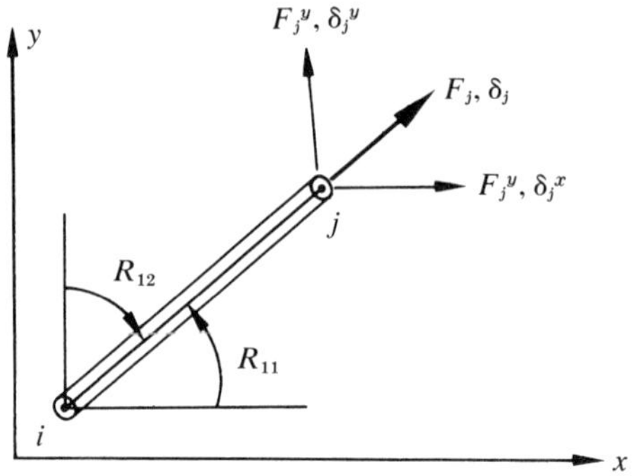

Fig. 6.4. Axial rod element in two dimensions.

While Eq. (6.9) reduces to the following,

$$\sigma_r = [-h_1 \quad -h_2 \quad h_1 \quad h_2] \begin{bmatrix} \delta_i^x \\ \delta_i^y \\ \delta_j^x \\ \delta_j^y \end{bmatrix} \tag{6.13}$$

Equation (6.11) in conjunction with Eqs. (6.2) and (6.9) (space truss) or Eqs. (6.12) and (6.13) (coplanar truss) is the required equation upon which the algorithm of nonlinear analysis is built. This algorithm and the computer procedure are presented in the next section.

6.3 ALGORITHM FOR NONLINEAR ANALYSIS

The nonlinear method of analysis, as presented here, is based on a relaxation procedure using successive linearly elastic approximations to obtain the nonlinear solution. The algorithm of the method can best be illustrated by considering the truss-structure shown in Fig. 6.5. For the purpose of simplicity, all rods will be assumed to conform to the stress-strain curve shown in Fig. 6.6.

If the truss is acted upon by a force, P, at node 1, as shown in Fig. 6.5 and the induced stress in any rod does not exceed the proportional limit of the material, the stresses in all rods and node 1 deflection can be determined by the linear elastic stiffness matrix method presented in Chapter 3. For the case in which P is of sufficient magnitude to cause the stress in rod 2, for instance, to exceed the proportional limit of the material, while the stresses in rods 1 and 3 are below the proportional limit, the results obtained using the linear elastic stiffness matrix method would be erroneous. However, the result thus obtained will be used as a first approxima-

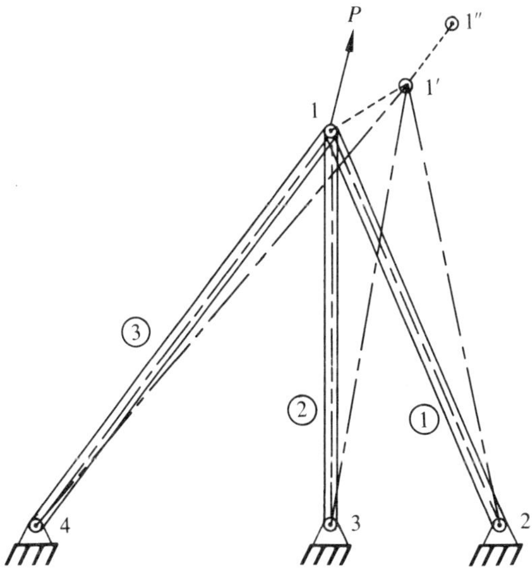

Fig. 6.5. Coplanar pin-ended truss.

tion to the actual state of stress in the truss for the given external applied force.

The results of linear analysis would show that the truss has undergone a deformation such that node 1 has moved to a new position 1′. However, 1′ would be the new position of node 1 only if the stresses in rods 1, 2, and 3 have not exceeded the proportional limit. The stress in rod 2 for the linear elastic analysis would equal OA and the corresponding strain would be

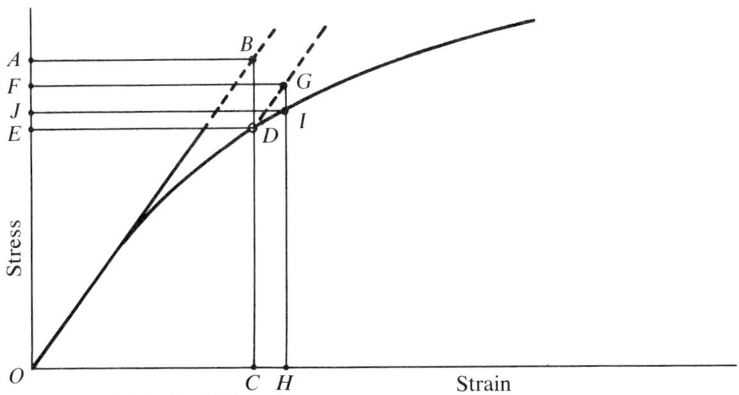

Fig. 6.6. Stress-strain curve of material for the truss in Fig. 6.5.

OC, locating point B in Fig. 6.6. The point B, as can be seen, does not lie on the actual stress-strain curve, thus it is impossible for the stress, OA, and the corresponding strain, OC, to exist in rod 2. Therefore, to hold node 1 in position 1′, an additional force is required at node 1 in the direction 1 to 3. Such a force, will be referred to as a fictitious force which we temporarily introduce as an external applied force. The magnitude of the fictitious force is equal to the difference between the force in rod 2 as computed by the linear elastic method and the true force in rod 2 for position 1′. When node 1 is in position 1′, the true stress and strain in rod 2 are OE and OC, respectively, as given by point D. Therefore, the magnitude of the fictitious force required to hold node 1 in position 1′ is equal to the product of the stress EA and the cross-sectional area of rod 2.

The fictitious force does not actually exist in the truss; therefore, it must be eliminated by applying another force of equal magnitude but in the opposite direction of the fictitious force. By applying this force at node 1 in the direction 3 to 1 and using the linear elastic method, node 1 will deflect from 1′ to a new position 1″ and additional stresses will be induced in all three rods. It will again be assumed that only the stress in rod 2 is above the proportional limit. The stress in rod 2 will load up from point D to point G where the line DG is parallel to OB. Once again, since the point G does not lie on the true stress-strain curve, OF and OH cannot represent the actual stress and strain, respectively, in rod 2.

For node 1 to remain in the new position 1″, the magnitude of the required fictitious force is equal to the product of the stress JF and the cross-sectional area of rod 2; its direction is 1 to 3. The fictitious force is eliminated by applying another force at node 1 of equal magnitude, but in the opposite direction of the fictitious force. This will cause node 1 to deflect to a new position and additional stresses will result in the three rods for a third approximation. The process can be continued until the actual state of stress in the truss is determined to any desired degree of accuracy. The method will converge at a rate dependent upon both the degree of accuracy desired and how far above the proportional limit are the stresses in the rods. Figure 6.7 shows a flow chart of the procedure as done on a digital computer.

Since the method is an iterative process it is necessary to set up a tolerance for convergence. This tolerance is the maximum deviation allowed between the linear stress and the true stress, for instance the stress increment GI in Fig. 6.6. Predetermined points on the actual stress-strain curve are used as the pivot points for iterations.

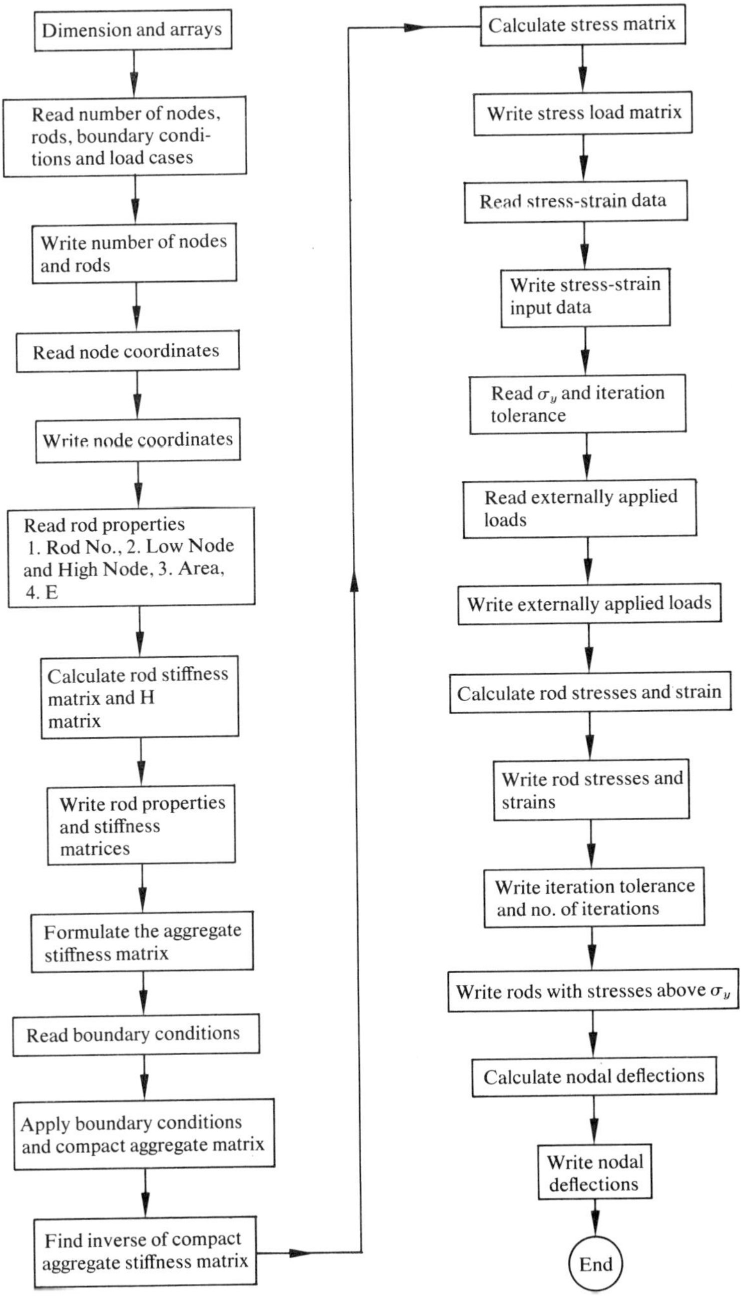

Fig. 6.7. Flow chart.

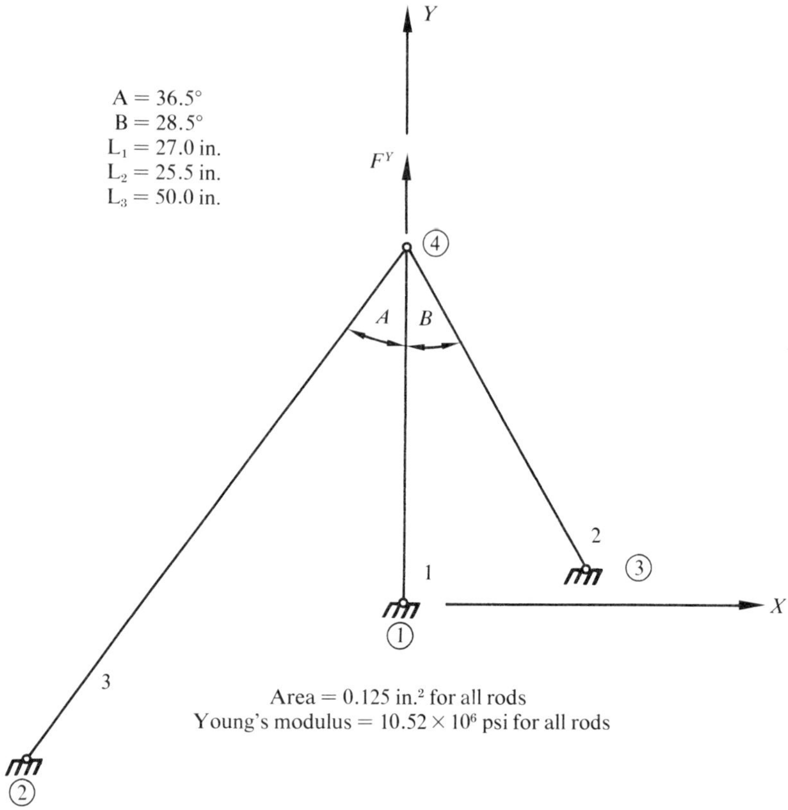

Fig. 6.8. Example 1 truss.

Illustrative Examples

Example 1

The first truss example to be analyzed, shown in Fig. 6.8, was originally studied both experimentally and analytically by Steinbacher *et al.* [25]. The truss is coplanar and consists of pin-ended axial rods having the stress-strain curve as shown in Fig. 6.9. Computer output for rod stresses is given in Table 1. Figures 6.10 and 6.11 show the results of the present non-linear analysis in comparison with those of Denke [4] and Steinbacher [25].

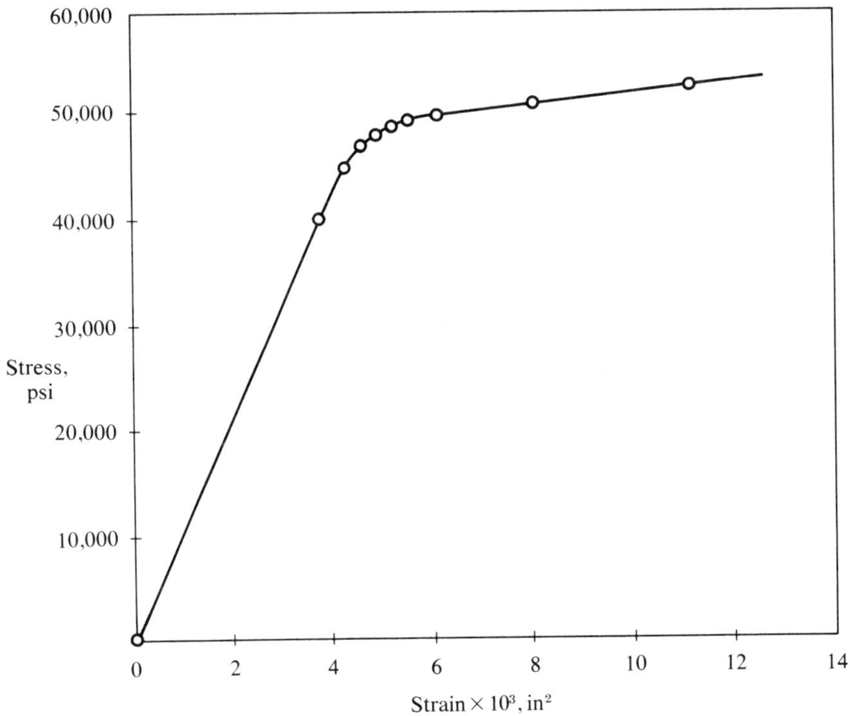

Fig. 6.9. Stress-strain curve for 24S-T4 aluminum alloy.

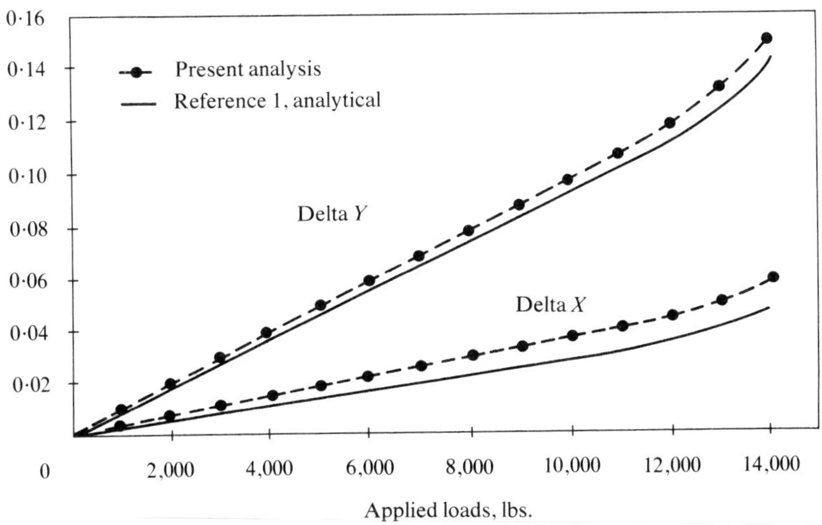

Fig. 6.10. Deflections of node 4 versus applied load for example 1.

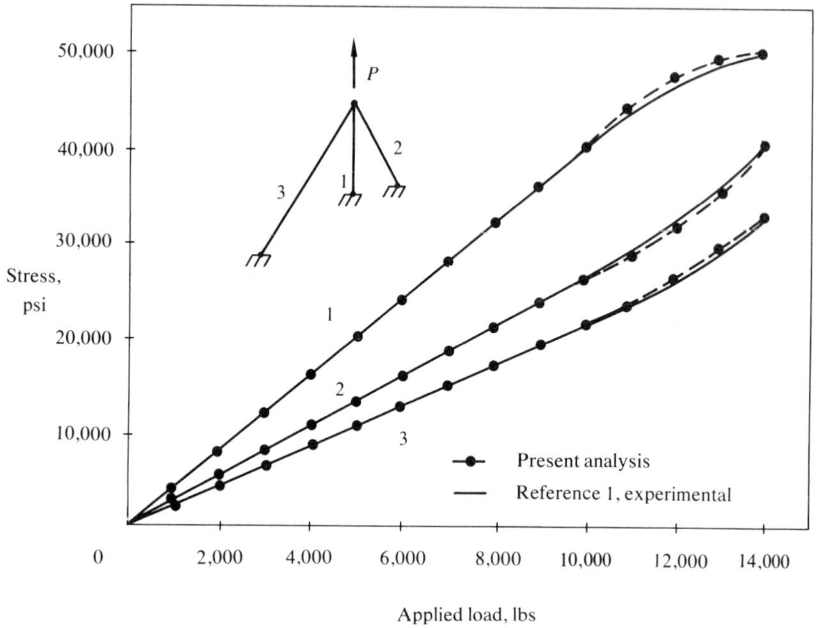

Fig. 6.11. Stress in the rods versus applied load for example 1.

Table 1. Rod stresses of Example 1.

Applied Load (lbs.)	Stresses (psi)		
	Rod 1	Rod 2	Rod 3
1,000	4,000	2,630	2,110
2,000	8,000	5,250	4,210
3,000	12,000	7,880	6,320
4,000	16,000	10,500	8,430
5,000	20,000	13,130	10,530
6,000	24,000	15,750	12,640
7,000	27,990	18,380	14,740
8,000	31,990	21,010	16,850
9,000	35,990	23,630	18,960
10,000	39,990	26,260	21,060
11,000	43,990	28,880	23,170
12,000	47,230*	31,950	25,630
13,000	49,090*	35,980	28,870
14,000	49,850*	40,750	32,690

*means stresses above proportional limit.

Example 2

The second illustrative truss example is shown in Fig. 6.12. All rods of the truss are considered to be made of a mild structural steel having the stress-strain curve shown in Fig. 6.13.

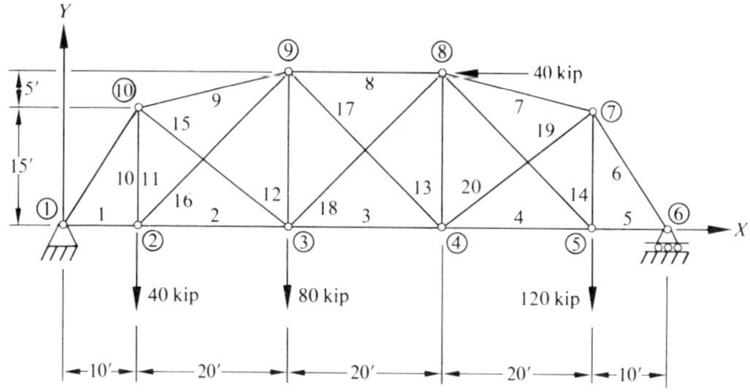

Rods 1–10, area = 4 in²

Rods 11–20, area = 2 in²

Fig. 6.12. Example 2 truss.

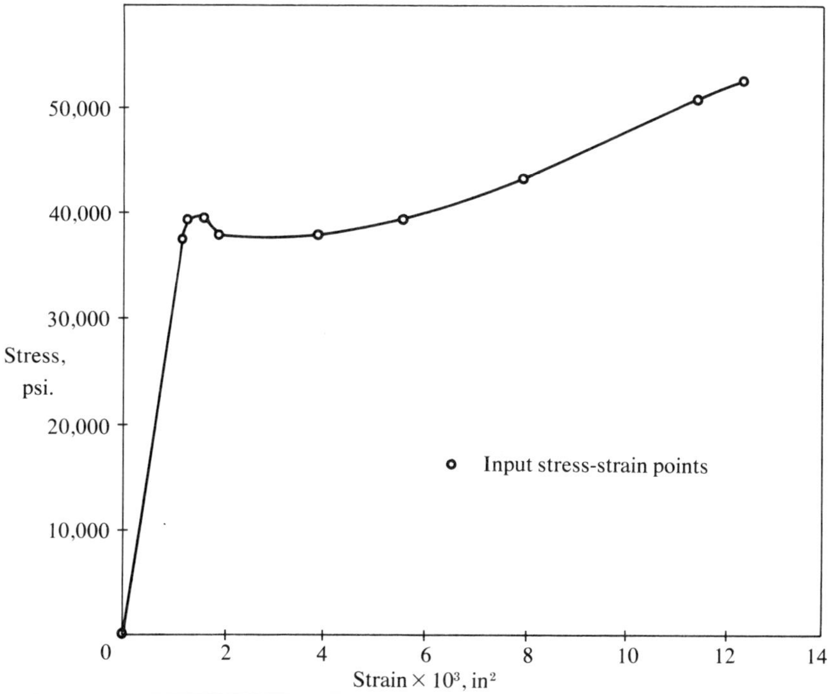

Fig. 6.13. Stress-strain curve for mild structural steel.

Computer output for rod stresses and node point deflections is given in Table 2. Only stresses in rods 6 and 14 are above the proportional limit. However, the node point deflections along the bottom of the truss are affected considerably, as shown in Fig. 6.14. The linear deflections are calculated by the normal direct stiffness matrix method as presented in Chapter 3 and the nonlinear deflections are based on the modified stiffness method as presented in this chapter.

Table 2. Rod stresses and node deflections of Example 2.

Rod	Stress (psi)
1	8333
2	11707
3	19715
4	11611
5	21667
6	−39060
7	−29721
8	−30287
9	−28734
10	−33050
11	26746
12	22615
13	−8820
14	39250
15	23858
16	−9542
17	−2730
18	4348
19	28522
20	17918

	Deflection (inches)	
Node	Horizontal	Vertical
1	0	0
2	0.033	−0.746
3	0.127	−1.410
4	0.284	−1.266
5	0.377	−1.538
6	0.464	0
7	0.042	−0.644
8	0.122	−1.336
9	0.364	−1.230
10	0.448	−0.585

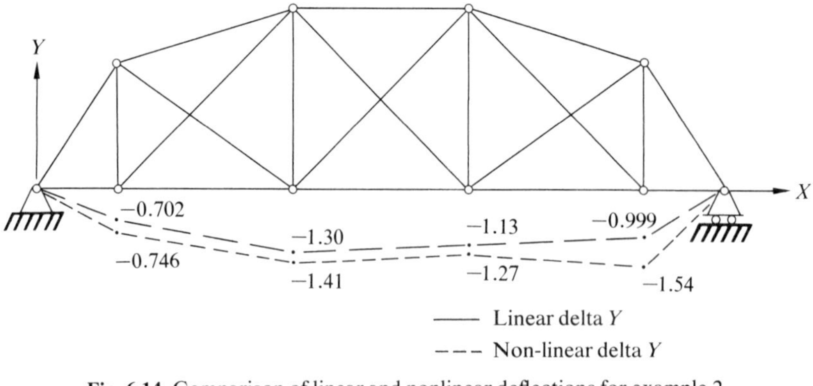

Fig. 6.14. Comparison of linear and nonlinear deflections for example 2.

CHAPTER SEVEN

Analysis by Structural Partitioning

7.1 INTRODUCTION

Quite often the structural analyst is faced with structural problems which if analyzed as a whole will require computer storage which may exceed the available computer capacity. If and when such large-scale problems arise, the structure may be partitioned (divided) into smaller structures such that each substructure is small enough that it will fit the available computer core storage. Assuming that the substructure-boundary conditions (displacements or loads) are known, then each substructure can be analyzed separately. Once all substructures are analyzed, then they are assembled back together by implementing conditions of continuity at common boundaries. This chapter presents the general theory of analysis by structural partitioning which was developed by Przemieniecki [20],* using the stiffness matrix method approach.

7.2 PROCEDURE OF ANALYSIS BY STRUCTURAL PARTITIONING USING THE STIFFNESS MATRIX METHOD

The analysis by structural partitioning is based on a relaxation procedure. Once the structure is partitioned into a discrete number of substructures (Fig. 7.1), then the common nodes of two adjacent substructures are considered to be, temporarily, fixed against any translational and rotational displacements. This will allow each substructure analysis to be

*Reproduced by permission of the American Institute of Aeronautics and Astronautics, from J. S. Przemieniecki's Matrix structural analysis of substructures in the *AIAA Journal*, 1963, **1**, 1, pp. 138–147.

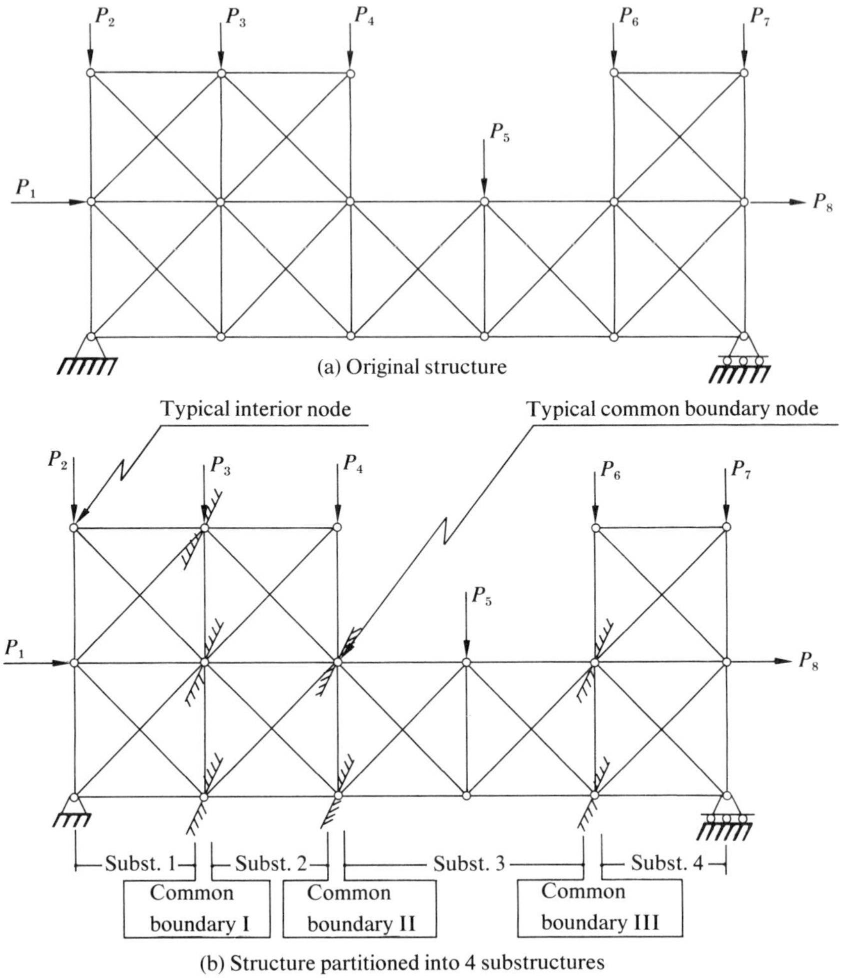

Fig. 7.1. Structure partitioning.

formulated and carried out separately, i.e., the displacements at the interior nodes can and will be determined. With the displacements known, the common boundary node reactions can be obtained and then combined with corresponding external nodal loads to form the total boundary loading. The common boundaries are, then, relaxed simultaneously and interior node displacements are calculated. The final actual displacements on the structure are determined by superposing the two displacements which are obtained first when the boundaries are in the fixed configuration and then the relaxed one. The internal loads are calculated from the structural element stiffness matrix equation in the same manner as was done in Chapter 3. Figure 7.2 shows a schematic of the procedural steps which are taken in the analysis by structural partitioning.

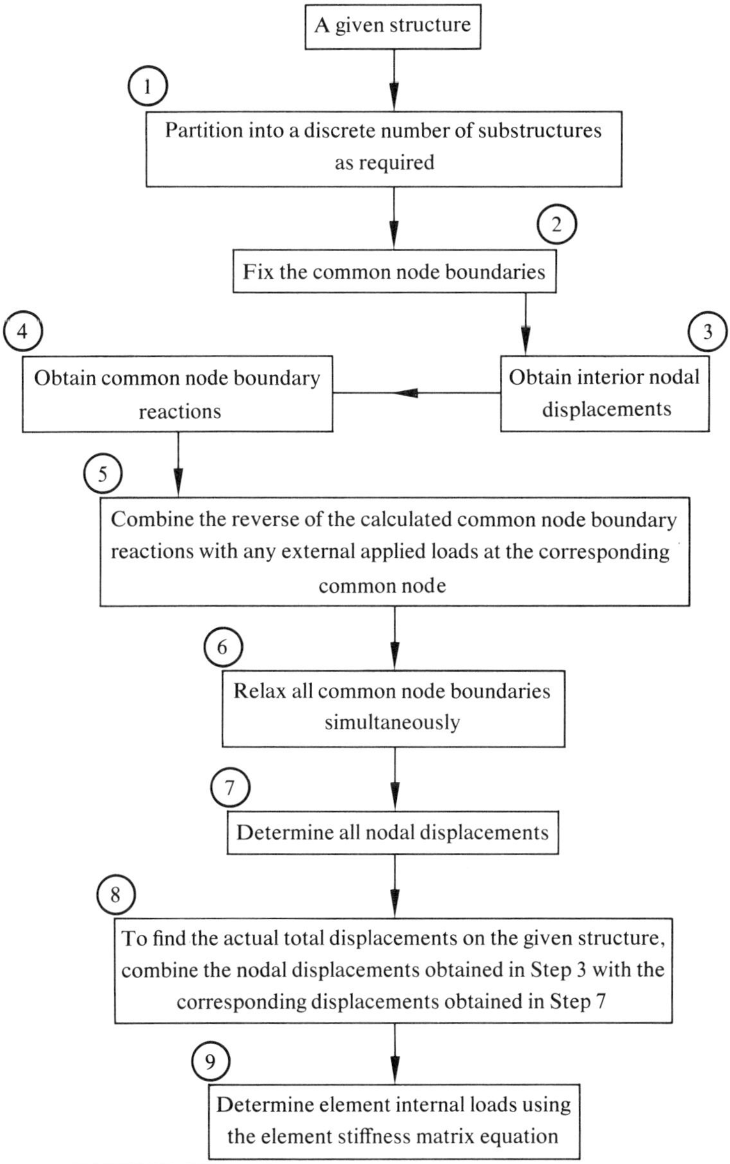

Fig. 7.2. Schematic for structural partitioning.

7.3 THEORY DEVELOPMENT FOR THE ANALYSIS BY STRUCTURAL PARTITIONING

In previous discussions, a structure was defined as an assemblage of structural elements interconnected at node points. In this chapter, for the purpose of analysis by structural partitioning, a structure is defined as an assemblage of substructures interconnected at common nodes. Regardless whether the structure is assumed to be composed of substructure assemblies or structural elements, the matrix equation format in regard to the functional relationship between forces and displacements is identical. Thus, considering the kth substructure, this relationship may be expressed as follows,

$$\{Q^k\} = [S^k]\{q^k\}, \tag{7.1}$$

where $[S^k]$ is the stiffness matrix for the kth substructure, $\{Q^k\}$ = external applied loads, and $\{q^k\}$ = nodal displacements.

Equation (7.1) can be partitioned in such a manner that the common boundary nodal forces and displacements are isolated from the interior nodal forces and displacements.

Thus, after partitioning, Eq. (7.1) takes on the form,

$$\left[\begin{array}{c}\{Q_c^k\} \\ \hline \{Q_i^k\}\end{array}\right] = \left[\begin{array}{c|c}[S_{cc}^k] & [S_{ci}^k] \\ \hline [S_{ic}^k] & [S_{ii}^k]\end{array}\right]\left[\begin{array}{c}\{q_c^k\} \\ \hline \{q_i^k\}\end{array}\right], \tag{7.2}$$

where the subscripts c and i denote the relationship of forces and displacements associated with common nodes on two adjacent substructures and interior nodes, respectively.

The column vectors of forces and displacements appearing in Eq. (7.2) can be represented as follows,

$$\left[\begin{array}{c}\{Q_c^k\} \\ \hline \{Q_i^k\}\end{array}\right] = \left[\begin{array}{c}\{Q_c^{k\varphi}\} \\ \hline \{Q_i^{k\varphi}\}\end{array}\right] + \left[\begin{array}{c}\{Q_c^{k\rho}\} \\ \hline \{Q_i^{k\rho}\}\end{array}\right] \tag{7.3}$$

$$\left[\begin{array}{c}\{q_c^k\} \\ \hline \{q_i^k\}\end{array}\right] = \left[\begin{array}{c}\{q_c^{k\varphi}\} \\ \hline \{q_i^{k\varphi}\}\end{array}\right] + \left[\begin{array}{c}\{q_c^{k\rho}\} \\ \hline \{q_i^{k\rho}\}\end{array}\right] \tag{7.4}$$

where the superscripts φ and ρ denote the conditions in which the common boundary nodes are fixed and relaxed, respectively.

7.4 SUBSTRUCTURE DISPLACEMENTS AND FORCES WITH COMMON BOUNDARY NODES FIXED

Upon fixing common boundary nodes against displacements and considering the kth substructure, it is obvious to recognize that the displacements and forces which appear in Eq. (7.2) become,

$$\{q_c^k\} = \{q_c^{k\varphi}\} = \{0\}, \tag{7.5a}$$

$$\{q_i^k\} = \{q_i^{k\varphi}\}, \tag{7.5b}$$

$$\{Q_c^k\} = \{Q_c^{k\varphi}\} = \{R_c^{k\varphi}\}, \tag{7.6a}$$

$$\{Q_i^k\} = \{Q_i^{k\varphi}\}, \tag{7.6b}$$

where $\{R_c^{k\varphi}\}$ is a column vector which represents the required restraint forces to maintain the zero displacement vector $\{q_c^{k\varphi}\}$, and $\{Q_i^k\}$ is a vector of interior nodal applied forces.

Note here, that a solution for both the boundary reactions $\{R_c^{k\varphi}\}$ and the interior nodal displacements $\{q_i^{k\varphi}\}$ is required before a final solution for the whole structure can be obtained.

If Eqs. (7.5) and (7.6) are substituted into Eq. (7.2) this will render the stiffness matrix $[S^k]$ nonsingular, and a solution for the interior node displacements $\{q_i^{k\varphi}\}$ and boundary reactions $\{R_c^{k\varphi}\}$ can be obtained. Thus, carrying out the substitution yields,

$$\begin{bmatrix} \{R_c^{k\varphi}\} \\ \hline \{Q_i^k\} \end{bmatrix} = \begin{bmatrix} [S_{cc}^k] & [S_{ci}^k] \\ \hline [S_{ic}^k] & [S_{ii}^k] \end{bmatrix} \begin{bmatrix} \{0\} \\ \hline \{q_i^{k\varphi}\} \end{bmatrix} \tag{7.7}$$

Solving Eq. (7.7) for the interior nodal displacements and common boundary node reactions, the following are obtained,

$$\{q_i^{k\varphi}\} = [S_{ii}^k]^{-1}\{Q_i^k\}, \tag{7.8}$$

and

$$\{R_c^{k\varphi}\} = [S_{ci}^k][S_{ii}^k]^{-1}\{Q_i^k\}. \tag{7.9}$$

7.5 SUBSTRUCTURE DISPLACEMENTS AND FORCES WITH COMMON BOUNDARY NODES RELAXED

Upon relaxation of common boundary nodes, the vector of displacements and forces which appear in Eq. (7.2) becomes,

$$\{Q_c^{k\rho}\} = \{Q_c^k\} - \{R_c^{k\varphi}\}, \tag{7.10a}$$

$$\{Q_i^k\} = \{Q_i^{k\rho}\} = \{0\}, \tag{7.10b}$$

$$\{q_c^k\} = \{q_c^{k\rho}\}, \tag{7.11a}$$

$$\{q_i^k\} = \{q_i^{k\rho}\}. \tag{7.11b}$$

Note that upon substitution of Eq. (7.9) into Eq. (7.10a), the following convenient form for $\{Q_c^{kp}\}$ is obtained,

$$\{Q_c^{kp}\} = \{Q_c^{k}\} - [S_{ci}^k][S_{ii}^k]^{-1}\{Q_i^k\}. \quad (7.12)$$

Utilizing Eqs. (7.10) and (7.11) in Eq. (7.2) yields,

$$\begin{bmatrix} \{Q_c^{kp}\} \\ \hline \{0\} \end{bmatrix} = \begin{bmatrix} [S_{cc}^k] & \vert & [S_{ci}^k] \\ \hline [S_{ic}^k] & \vert & [S_{ii}^k] \end{bmatrix} \begin{bmatrix} \{q_c^{kp}\} \\ \hline \{q_i^{kp}\} \end{bmatrix}. \quad (7.13)$$

From Eq. (7.13), it can be easily seen that the interior nodal displacements can be expressed in terms of the common boundary nodal displacement as follows,

$$\{q_i^{kp}\} = -[S_{ii}^k]^{-1}[S_{ic}]\{q_c^{kp}\}. \quad (7.14)$$

Also, from Eq. (7.13), the common boundary nodal forces are,

$$\{Q_c^{kp}\} = [S_{cc}^k]\{q_c^{kp}\} + [S_{ci}^k]\{q_i^{kp}\}. \quad (7.15)$$

Upon substitution of Eq. (7.14) into Eq. (7.15) the following functional relationship between the forces and corresponding displacements at the common boundaries is obtained,

$$\{Q_c^{kp}\} = [S_c^k]\{q_c^{kp}\}, \quad (7.16)$$

where

$$[S_c^k] = \Big[[S_{cc}^k] - [S_{ci}^k][S_{ii}^k]^{-1}[S_{ic}^k] \Big]. \quad (7.17)$$

The stiffness matrix $[S_c^k]$ represents the stiffness contribution of the kth substructure to the common boundary joining nodes only. This matrix will be referred to as the substructure reduced stiffness matrix.

In order to determine the common boundary displacements for the total structure, the stiffness contributions from all substructures must be superposed to yield,

$$\{Q_c^p\} = [S_c]\{q_c^p\}, \quad (7.18)$$

or

$$\{q_c^p\} = [S_c]^{-1}\{Q_c^p\}. \quad (7.19)$$

The matrix $[S_c]$ is the reduced aggregate stiffness matrix for the whole structure through which common boundary nodal displacements only are related to corresponding common boundary nodal applied forces. This matrix is formed by superposing all substructure matrices as shown in Fig. 7.3.

Likewise, the force vector $\{Q_c^p\}$ is formed by superposing all the substructure boundary force matrices $\{Q_c^{kp}\}$ as shown in Fig. 7.4.

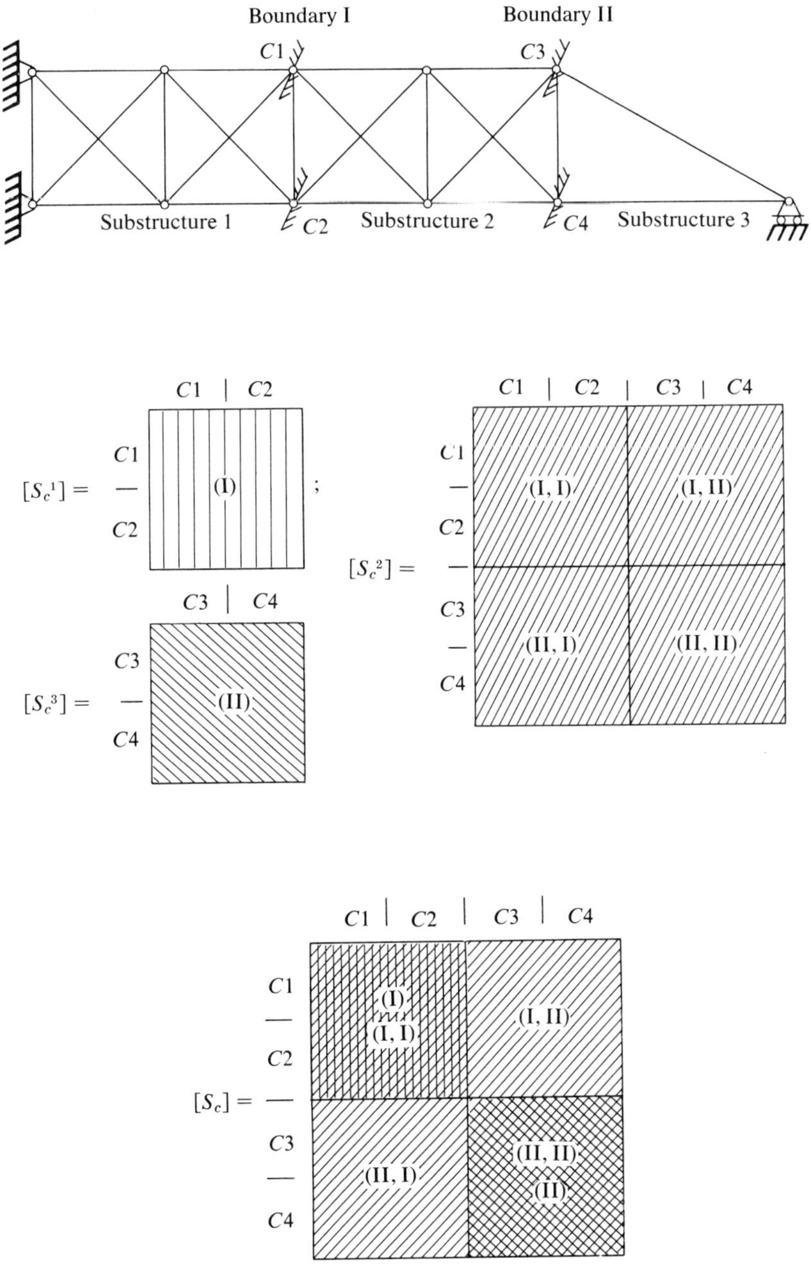

Fig. 7.3. Schematic for determining the common boundary reduced aggregate stiffness matrix.

7.6 STRUCTURE DISPLACEMENTS AND INTERNAL FORCES

The actual nodal displacements on the structure are given by Eq. (7.19) for the common boundary nodes, while for the interior nodes, the displacements are given by combining Eqs. (7.8) and (7.14), or,

$$\{q_i^k\} = [S_{ii}^k]^{-1}\{Q_i^k\} - [S_{ii}^k]^{-1}[S_{ic}]\{q_c^{kp}\}. \tag{7.20}$$

Having the nodal displacements, the internal forces are calculated by using the functional relationship between displacements and corresponding forces for each structural element. This is discussed in detail in Chapter 3.

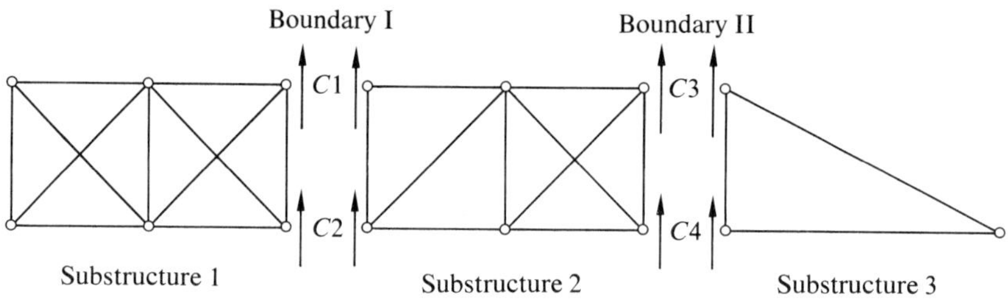

(a) Structure in partitioned form.

$$\{Q_c^{1p}\} = \begin{bmatrix} C1 \\ C2 \end{bmatrix}, \qquad \{Q_c^{2p}\} = \begin{bmatrix} C1 \\ C2 \\ C3 \\ C4 \end{bmatrix} \qquad \{Q_c^{3p}\} = \begin{bmatrix} C3 \\ C4 \end{bmatrix}$$

(b) Substructure common boundary force vectors,

$$\{Q_c^p\} = \Sigma_i\{Q_c^{ip}\} = \begin{bmatrix} C1 \\ C2 \\ 0 \\ 0 \end{bmatrix} + \begin{bmatrix} C1 \\ C2 \\ C3 \\ C4 \end{bmatrix} + \begin{bmatrix} 0 \\ 0 \\ C3 \\ C4 \end{bmatrix}$$

(c) Structure common boundary force vector Q_c^p.

Fig. 7.4. Schematic for determining the structure common boundary forces.

Illustrative Example

To illustrate the method of analysis by structural partitioning using the stiffness method, a planar-truss structure (Fig. 7.5) will be considered. For simplicity, all rods will be assumed to be made of the same material and have a modulus of elasticity equal to 10^7 psi. A cross-sectional area of 1.0 in^2 is assumed for each horizontal and vertical rod and $\sqrt{2}$ in^2 for all others.

We begin the analysis by partitioning the structure into a discrete number of substructures. In our present example, the truss will be partitioned

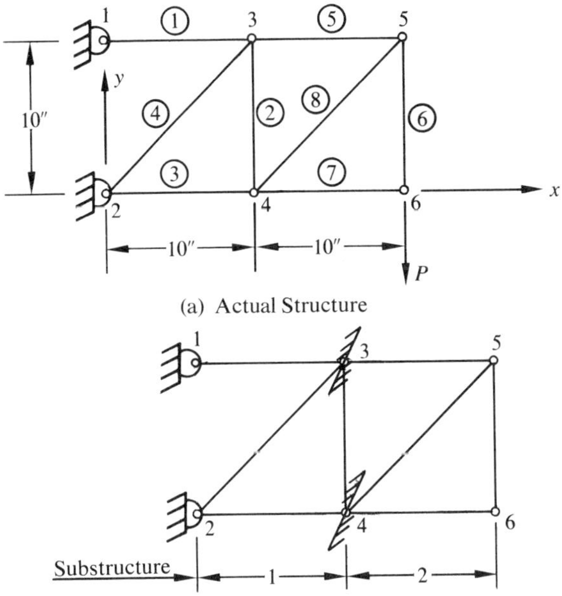

(a) Actual Structure

(b) Partitioned Structure

Fig. 7.5. Truss structure.

into two substructures by introducing temporary restraints at nodes 3 and 4 as shown below.

Having partitioned the structure, the next step in the analysis is to form the aggregate stiffness matrix for each substructure shown in Fig. 7.6. To do so, the stiffness matrix for each element in any one substructure must be obtained first. Thus, the stiffness matrices for the elements appearing in substructure 1 can be obtained easily by referring to Chapter 3 and are given in the following equations,

Rod Element 1

$$\begin{bmatrix} F_1{}^x \\ F_1{}^y \\ F_3{}^x \\ F_3{}^y \end{bmatrix} = 10^6 \begin{bmatrix} 1 & 0 & -1 & 0 \\ 0 & 0 & 0 & 0 \\ -1 & 0 & 1 & 0 \\ 0 & 0 & 0 & 0 \end{bmatrix} \begin{bmatrix} \delta_1{}^x \\ \delta_1{}^y \\ \delta_3{}^x \\ \delta_3{}^y \end{bmatrix} \tag{7.21a}$$

Rod Element 2

$$\begin{bmatrix} F_3{}^x \\ F_3{}^y \\ F_4{}^x \\ F_4{}^y \end{bmatrix} = 10^6 \begin{bmatrix} 0 & 0 & 0 & 0 \\ 0 & 1 & 0 & -1 \\ 0 & 0 & 0 & 0 \\ 0 & -1 & 0 & 1 \end{bmatrix} \begin{bmatrix} \delta_3{}^x \\ \delta_3{}^y \\ \delta_4{}^x \\ \delta_4{}^y \end{bmatrix} \tag{7.21b}$$

Rod Element 3

$$\begin{bmatrix} F_2{}^x \\ F_2{}^y \\ F_4{}^x \\ F_4{}^y \end{bmatrix} = 10^6 \begin{bmatrix} 1 & 0 & -1 & 0 \\ 0 & 0 & 0 & 0 \\ -1 & 0 & 1 & 0 \\ 0 & 0 & 0 & 0 \end{bmatrix} \begin{bmatrix} \delta_2{}^x \\ \delta_2{}^y \\ \delta_4{}^x \\ \delta_4{}^y \end{bmatrix} \tag{7.21c}$$

Rod Element 4

$$\begin{bmatrix} F_2{}^x \\ F_2{}^y \\ F_3{}^x \\ F_3{}^y \end{bmatrix} = 10^6 \begin{bmatrix} 0.5 \\ 0.5 \\ -0.5 \\ -0.5 \end{bmatrix} \begin{bmatrix} 0.5 & -0.5 & -0.5 \\ 0.5 & -0.5 & -0.5 \\ -0.5 & 0.5 & 0.5 \\ -0.5 & 0.5 & 0.5 \end{bmatrix} \begin{bmatrix} \delta_2{}^x \\ \delta_2{}^y \\ \delta_3{}^x \\ \delta_3{}^y \end{bmatrix} \tag{7.21d}$$

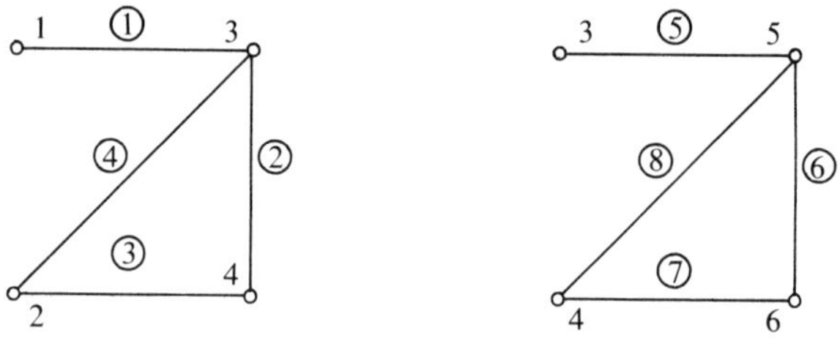

Fig. 7.6. Disassembled substructures.

Superposing the element stiffness matrices which are given in Eqs. (7.2a), (7.2b), (7.2c), and (7.2d), yield the following aggregate stiffness matrix for substructure 1,

$$[S^2] = 10^6 \begin{bmatrix} 1.0 & & & & -1.0 & & & \\ & & & & & & & \\ & & 1.5 & 0.5 & -0.5 & -0.5 & -1.0 & \\ & & 0.5 & 0.5 & -0.5 & -0.5 & & \\ -1.0 & & -0.5 & -0.5 & 1.5 & 0.5 & & \\ & & -0.5 & -0.5 & 0.5 & 1.5 & & -1.0 \\ & & -1.0 & & & & 1.0 & \\ & & & & & -1.0 & & 1.0 \end{bmatrix} \begin{matrix} \text{Node 1} \\ \\ \text{Node 2} \\ \\ \text{Node 3} \\ \\ \text{Node 4} \\ \\ \end{matrix} \tag{7.22}$$

(columns: Node 1, Node 2, Node 3, Node 4)

The stiffness coefficients which do not appear in Eq. (7.22) are zeros.

The stiffness matrix equations for the rod elements in substructure 2 are,

Rod Element 5

$$\begin{bmatrix} F_3^x \\ F_3^y \\ F_5^x \\ F_5^y \end{bmatrix} = 10^6 \begin{bmatrix} 1 & 0 & -1 & 0 \\ 0 & 0 & 0 & 0 \\ -1 & 0 & 1 & 0 \\ 0 & 0 & 0 & 0 \end{bmatrix} \begin{bmatrix} \delta_3^x \\ \delta_3^y \\ \delta_5^x \\ \delta_5^y \end{bmatrix} \tag{7.23a}$$

Rod Element 6

$$\begin{bmatrix} F_5^x \\ F_5^y \\ F_6^x \\ F_6^y \end{bmatrix} = 10^6 \begin{bmatrix} 0 & 0 & 0 & 0 \\ 0 & 1 & 0 & -1 \\ 0 & 0 & 0 & 0 \\ 0 & -1 & 0 & 1 \end{bmatrix} \begin{bmatrix} \delta_5^x \\ \delta_5^y \\ \delta_6^x \\ \delta_6^y \end{bmatrix} \tag{7.23b}$$

Rod Element 7

$$\begin{bmatrix} F_4^x \\ F_4^y \\ F_6^x \\ F_6^y \end{bmatrix} = 10^6 \begin{bmatrix} 1 & 0 & -1 & 0 \\ 0 & 0 & 0 & 0 \\ -1 & 0 & 1 & 0 \\ 0 & 0 & 0 & 0 \end{bmatrix} \begin{bmatrix} \delta_4^x \\ \delta_4^y \\ \delta_6^x \\ \delta_6^y \end{bmatrix} \tag{7.23c}$$

Rod Element 8

$$\begin{bmatrix} F_4^x \\ F_4^y \\ F_5^x \\ F_5^y \end{bmatrix} = 10^6 \begin{bmatrix} 0.5 & 0.5 & -0.5 & -0.5 \\ 0.5 & 0.5 & -0.5 & -0.5 \\ -0.5 & -0.5 & 0.5 & 0.5 \\ -0.5 & -0.5 & 0.5 & 0.5 \end{bmatrix} \begin{bmatrix} \delta_4^x \\ \delta_4^y \\ \delta_5^x \\ \delta_5^y \end{bmatrix} \tag{7.23d}$$

Superposing the element stiffness matrices which appear in Eqs. (7.23a), (7.23b), (7.23c), and (7.23d), yields the following aggregate stiffness matrix for substracture 2,

$$[S^2] = 10^6 \begin{array}{c} \begin{array}{cccc} \text{Node 3} & \text{Node 4} & \text{Node 5} & \text{Node 6} \end{array} \\ \left[\begin{array}{cccccccc} 1.0 & & & & -1.0 & & & \\ & & & & & & & \\ & & 1.5 & 0.5 & -0.5 & -0.5 & -1.0 & \\ & & 0.5 & 0.5 & -0.5 & -0.5 & & \\ -1.0 & & -0.5 & -0.5 & 1.5 & 0.5 & & \\ & & -0.5 & -0.5 & 0.5 & 1.5 & & -1.0 \\ & & -1.0 & & & & 1.0 & \\ & & & & & -1.0 & & 1.0 \end{array}\right] \end{array} \begin{array}{l} \text{Node 3} \\ \\ \text{Node 4} \\ \\ \text{Node 5} \\ \\ \text{Node 6} \\ \\ \end{array} \tag{7.24}$$

To form the aggregate common boundary stiffness matrices, Eq. (7.17) will be used. Note that nodes 5 and 6 are the only interior nodes on substructure 2, while on substructure 1 there are no interior nodes. Thus for the later substructure, the common boundary reduced aggregate stiffness matrix is,

$$[S_c^1] = 10^6 \begin{bmatrix} 1.5 & 0.5 & 0 & 0 \\ 0.5 & 1.5 & 0 & -1.0 \\ 0 & 0 & 1.0 & 0 \\ 0 & -1.0 & 0 & 1.0 \end{bmatrix} \tag{7.25}$$

For substructure 2, the common boundary reduced aggregate stiffness matrix is,

$$[S_c^2] = [S_{cc}^2] - [S_{ci}^2][S_{ii}^2]^{-1}[S_{ic}^2]$$

$$= 10^6 \begin{bmatrix} 1 & 0 & 0 & 0 \\ 0 & 0 & 0 & 0 \\ 0 & 0 & 1.5 & 0.5 \\ 0 & 0 & 0.5 & 0.5 \end{bmatrix} - 10^6 \begin{bmatrix} -1 & 0 & 0 & 0 \\ 0 & 0 & 0 & 0 \\ -0.5 & -0.5 & -1 & 0 \\ -0.5 & -0.5 & 0 & 0 \end{bmatrix}$$

$$\times \begin{bmatrix} 15(10)^5 & 5(10)^5 & 0 & 0 \\ 5(10)^5 & 15(10)^5 & 0 & -10^6 \\ 0 & 0 & 10^6 & 0 \\ 0 & -10^6 & 0 & 10^6 \end{bmatrix}^{-1}$$

$$\times \begin{bmatrix} -1 & 0 & -0.5 & -0.5 \\ 0 & 0 & -0.5 & -0.5 \\ 0 & 0 & -1 & 0 \\ 0 & 0 & 0 & 0 \end{bmatrix} 10^6 = \begin{bmatrix} 0 & 0 & 0 & 0 \\ 0 & 0 & 0 & 0 \\ 0 & 0 & 0 & 0 \\ 0 & 0 & 0 & 0 \end{bmatrix} \tag{7.26}$$

To form the common boundary reduced aggregate stiffness matrix for the entire structure, we combine the stiffness matrices in Eqs. (7.25) and (7.26).

$$[S_c] = [S_c^1] + [S_c^2] = 10^6 \begin{bmatrix} 1.5 & 0.5 & 0 & 0 \\ 0.5 & 1.5 & 0 & -1 \\ 0 & 0 & 1 & 0 \\ 0 & -1 & 0 & 1 \end{bmatrix} \tag{7.27}$$

Since no interior nodes exist in substructure 1, its common boundary forces are zero. Hence, the common boundary forces for the entire structure are (*see* Eq. 7.12).

$$\{Q_c^p\} = \{Q_c^{2p}\} = \{Q_c^2\} - [S_{ci}^2][S_{ii}^2]^{-1}\{Q_i^2\}$$

$$= \begin{bmatrix} 0 \\ 0 \\ 0 \\ -P \end{bmatrix} - 10^6 \begin{bmatrix} -1 & 0 & 0 & 0 \\ 0 & 0 & 0 & 0 \\ -0.5 & -0.5 & -1 & 0 \\ -0.5 & -0.5 & 0 & 0 \end{bmatrix} \begin{bmatrix} 15(10)^5 & 5(10)^5 & 0 & 0 \\ 5(10)^5 & 15(10)^5 & 0 & -10^6 \\ 0 & 0 & 10^6 & 0 \\ 0 & -10^6 & 0 & 10^6 \end{bmatrix} \begin{bmatrix} 0 \\ 0 \\ 0 \\ -P \end{bmatrix}$$

$$= \begin{bmatrix} P \\ 0 \\ -P \\ -2P \end{bmatrix} \tag{7.28}$$

From Eq. (7.19), the structure common boundary displacements are,

$$\{q_c^p\} = [S_c]^{-1}\{Q_c^p\}$$

$$= \begin{bmatrix} 15(10)^5 & 5(10)^5 & 0 & 0 \\ 5(10)^5 & 15(10)^5 & 0 & -10^6 \\ 0 & 0 & 10^6 & 0 \\ 0 & -10^6 & 0 & 10^6 \end{bmatrix} \begin{bmatrix} P \\ 0 \\ -P \\ -2P \end{bmatrix} = 10^{-6} \begin{bmatrix} 3P \\ -7P \\ -P \\ -9P \end{bmatrix} = \begin{bmatrix} \delta_3^x \\ \delta_3^y \\ \delta_4^x \\ \delta_4^y \end{bmatrix} \tag{7.29}$$

From Eq. (7.20) the interior nodal displacements for the structure are,

$$\{q_i\} = \{q_i^2\} = [S_{ii}^2]^{-1}\{Q_i^2\} - [S_{ii}^2]^{-1}[S_{ic}^2]\{q_c^2\}$$

$$= \begin{bmatrix} 15(10)^5 & 5(10)^5 & 0 & 0 \\ 5(10)^5 & 15(10)^5 & 0 & -10^6 \\ 0 & 0 & 10^6 & 0 \\ 0 & -10^6 & 0 & 10^6 \end{bmatrix}^{-1} \begin{bmatrix} 0 \\ 0 \\ 0 \\ -P \end{bmatrix}$$

$$- \begin{bmatrix} 15(10)^5 & 5(10)^5 & 0 & 0 \\ 5(10)^5 & 15(10)^5 & 0 & -10^6 \\ 0 & 0 & 10^6 & 0 \\ 0 & -10^6 & 0 & 10^6 \end{bmatrix}^{-1}$$

$$\times 10^6 \begin{bmatrix} -1 & 0 & -0.5 & -0.5 \\ 0 & 0 & -0.5 & -0.5 \\ 0 & 0 & -1 & 0 \\ 0 & 0 & 0 & 0 \end{bmatrix} \begin{bmatrix} 3P \\ -7P \\ -P \\ -9P \end{bmatrix} \times 10^{-6}$$

$$= 10^{-6} \begin{bmatrix} 4P \\ -16P \\ -P \\ -17P \end{bmatrix} = \begin{bmatrix} \delta_5^x \\ \delta_5^y \\ \delta_6^x \\ \delta_6^y \end{bmatrix} \tag{7.30}$$

With all the displacements known [Eqs. (7.29) and (7.30)], the internal forces on all structural elements may be obtained by utilizing Eqs. (7.21) and (7.23).

7.7 SUPPLEMENTARY PROBLEMS

In the following problems solve for nodal displacements and internal forces by employing the method of structural partitioning using the stiffness matrix approach.

Problem 7.1. The structure is a two-bays planar truss with externally applied loads as shown.

$E = 10^7$ psi
$A = 1.0\ \text{in}^2$ (horizontal and vertical members)
$A = \sqrt{2}\ \text{in}^2$ (diagonal members)

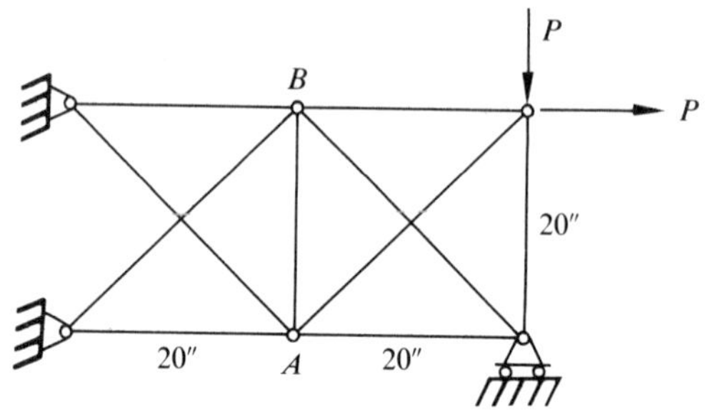

Problem 7.2. Same as Problem 7.1 with the addition of externally applied loads at $A(P, P)$ and $B(P, P)$.

Problem 7.3. The structure is a three-bays planar truss loaded as shown
$E = 10^7$ psi
$A = 1.0\ \text{in}^2$ (horizontal and vertical members)
$A = \sqrt{2}\ \text{in}^2$ (diagonal members)

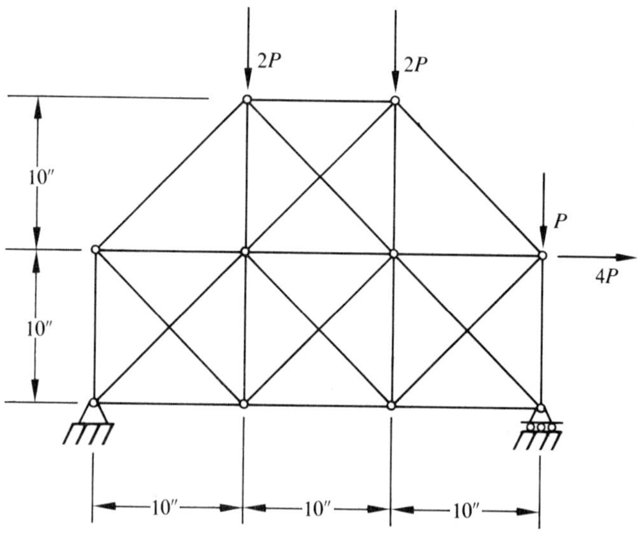

Problem 7.4. The structure is a truss which is given in Problem 3.4.

Problem 7.5. The structure is a rigid frame.
$E = 30 \times 10^6$ psi
$I = 1\ \text{in}^4$
$A = 1\ \text{in}^2$

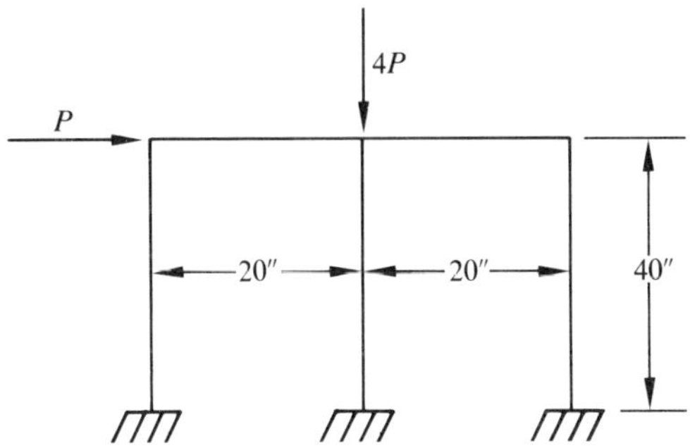

Problem 7.6. A rigid frame with a truss-roof.

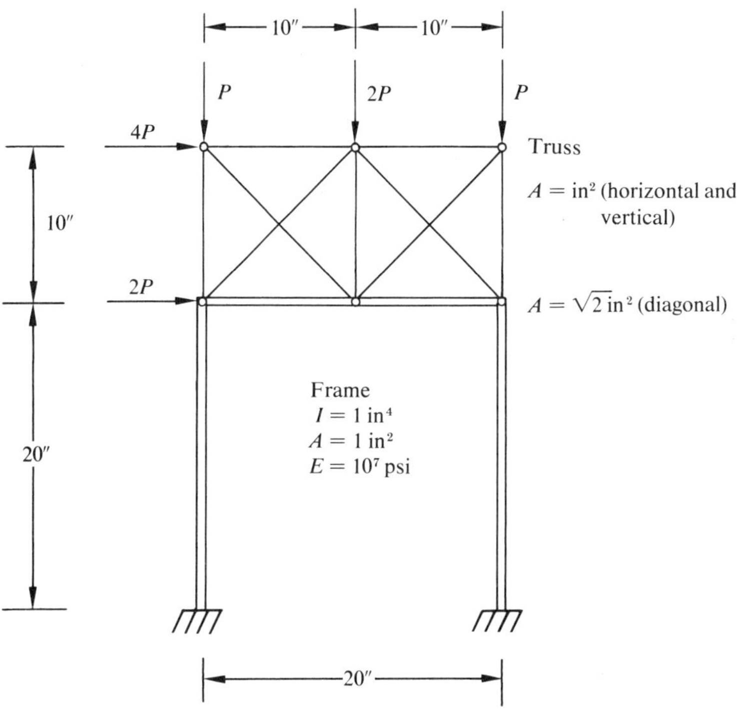

Problem 7.7. The structure is a beam which consists of shear panels and reinforcing axial rods.

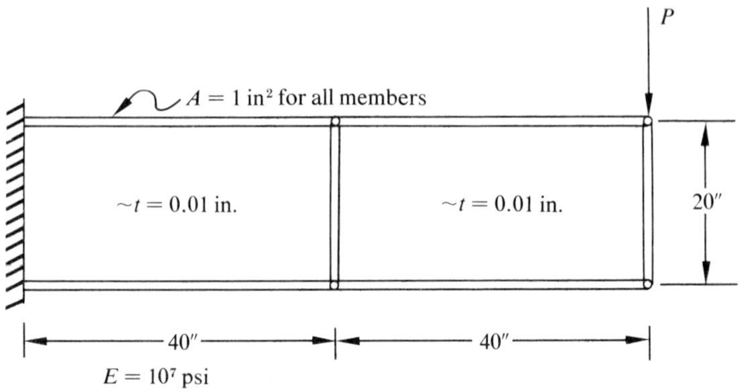

Problem 7.8. Same as Problem 7.7 except the beam is tapered as shown.

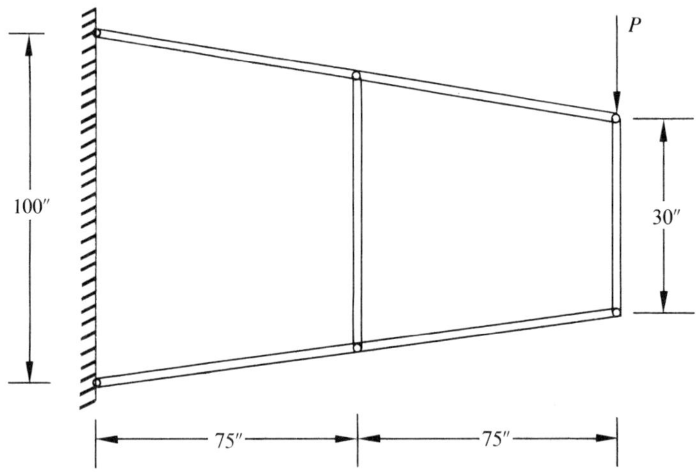

7.8 SUGGESTED COMPUTER EXERCISES

1. Write up a program to solve Problem 7.4. Assume your own input data.
2. Write a program to solve Problem 7.5. Assume your own input data.
3. Write a general purpose program for the analysis by structural partitioning.

References

1. Argyris, J. H., *Recent Advances in Matrix Methods of Structural Analysis*, Pergamon Press, 1964.
2. Benscoter, S. V., The Partitioning of Matrices in Structural Analysis, *Journal of Applied Mechanics*, *Trans. ASME*, **70**, pp. 303–307, 1948.
3. Benscoter, S. V., The Partitioning of Matrices in Structural Analysis, *J. A. Mech.* **15**, 1948.
4. Denke, P. H., A Matrix Solution of Certain Non-Linear Problems in Structural Analysis, *Journal of the Aeronautical Sciences*, **23**, pp. 231–236, March 1956.
5. Fraeijs de Veubeke, B., *Matrix Methods of Structural Analysis*, John Wiley, New York, 1965.
6. Gallagher, R. H., *A Correlation Study of Methods of Matrix Structural Analysis*, Report to the 14th Meeting, Structures and Materials Panel, Advisory Group for Aeronautical Research and Development, NATO, Paris, France, July 6, 1962.
7. Gallagher, R. H., Padlog, J. and Bijlaard, P. P., Stress Analysis of Heated Complex Shapes, *ARS Journal*, May 1962.
8. Gere, J. M., and W. Weaver, Jr., *Analysis of Framed Structures*, Van Nostrand, Princeton, New Jersey, 1965.
9. Goldberg, John E., and Ralph M. Richard, Analysis of Non-Linear Structures, *Proceedings ASCE Third Conference on Electronic Computation*, pp. 333–351, 1963.
10. Hensley, R. C., and Jamal J. Azar, Computer Analysis of Non-Linear Truss-Structures, *Journal of the Structural Engi. Div.*, ASCE **94**(ST6), 1968.
11. Jensen, W. R., W. E. Falby and N. Prince, Matrix Analysis Methods for Anisotropic Inelastic Structures, AFDL-TR-65-220, April 1966.
12. Kosko, E., Matrix Inversion by Partitioning, *The Aeronautical Quarterly*, **8**, pp. 157–184, May 1957.
13. Laursen, H. I., *Matrix Analysis of Structures*, The Macmillan Company, New York, 1962.
14. Livesley, R. K., *Matrix Methods of Structural Analysis*, Pergamon Press, Oxford, England, 1964.
15. Mallett, R. H., and P. V. Marcal, Finite Element Analysis of Nonlinear Structures, *Journal of the Structural Division, Proc.* ASCE, **94**, No. ST9, Sept. 1968, pp. 2081–2105.

16. Martin, H. C., *Introduction to Matrix Methods of Structural Analysis*, McGraw-Hill, New York, 1966.
17. Poppleton, E. D., Note on the Matrix Analysis of Non-Linear Structures, UTIA Technical Note No. 46, March 1961.
18. Przemieniecki, J. S., *Theory of Matrix Structural Analysis*, McGraw-Hill, New York, 1968.
19. Przemieniecki, J. S., *et al.* (Eds.), *Proceedings of Conference on Matrix Methods in Structural Mechanics*, AFFDL TR 66-80, Air Force Flight Dynamics Laboratory, October 1965.
20. Przemieniecki, J. S., Matrix Structural Analysis of Substructures, *AIAA Journal* 1963, **1**, No. 1, pp. 138–147.
21. Remberg, W., and W. R. Osgood. Description of Stress-Strain Curves by Three Parameters, NACA TN 902, July 1943.
22. Rowman, W. H., and R. M. Hackett, (Eds.), *Proceedings of the Symposium on Application of Finite Element Methods in Civil Engineering*, Vanderbilt University, Nashville, Tenn., 1969.
23. Rubinstein, M. F., *Matrix Computer Analysis of Structures*, Prentice-Hall, Englewood, New Jersey, 1966.
24. Spillers, W. R., *Automated Structural Analysis: An Introduction*, Pergamon Press, Inc., New York, 1971.
25. Steinbacher, F. R., C. N. Gaylord and W. K. Rey, Method for Analyzing Indeterminate Structures Stressed Above Proportional Limit, NACA TN 2376, June 1951.
26. Turner, M. J., R. W. Clough, H. C. Martin and L. J. Topp, Stiffness and Deflection Analysis of Complex Structures, *Journal of the Aeronautical Sciences*, **23**, pp. 805–823, September 1956.
27. Willems, N., and Lucas, W. M., Jr., *Matrix Analysis for Structural Engineers*, Prentice-Hall, Englewood Cliffs, New Jersey, 1968.
28. Wilson, E. L., Matrix Analysis of Non-Linear Structures, *Proceedings ASCE Second Conference on Electronic Computation*, pp. 415–428, 1960.
29. Zienkiewicz, O. C., *The Finite Element Method in Structural and Continuum Mechanics*, McGraw-Hill, New York, 1967.

Index